TV

LAND MOBILE RADIO SYSTEMS
Second Edition

LAND MOBILE RADIO SYSTEMS
Second Edition

Edward N. Singer

PTR Prentice Hall
Englewood Cliffs, NJ 07632

Library of Congress Cataloging-In-Publication Data

Singer, Edward N.
    Land mobile radio systems / Edward N. Singer.—second edition
       p. cm.
    Includes bibliographical references and index.
    ISBN 0-13-123159-6
    1. Mobile communications systems. I. Title.
TK6570.M6S56 1994
321.3845—dc20
                                      93-45021
                                      CIP

Editorial/production supervision: BooksCraft, Inc., Indianapolis, IN
Cover design: Jeanette Jacobs
Cover photo: The Stock Market, Bruno © 1993
Acquisitions editor: Karen Gettman
Manufacturing manager: Alexis R. Heydt

© 1994 by PTR Prentice Hall
Prentice-Hall, Inc.
A Paramount Communications Company
Englewood Cliffs, NJ 07632

The publisher offers discounts on this book when ordered
in bulk quantities. For more information contact:

       Corporate Sales Department
       PTR Prentice Hall
       113 Sylvan Avenue
       Englewood Cliffs, NJ 07632
       Phone: 201-592-2863
       FAX: 201-592-2249

Printed in the United States of America
10 9 8 7 6 5 4 3 2 1

ISBN 0-13-123159-6

Prentice-Hall International (UK) Limited, *London*
Prentice-Hall of Australia Pty. Limited, *Sydney*
Prentice-Hall Canada Inc., *Toronto*
Prentice-Hall Hispanoamericana, S.A., *Mexico*
Prentice-Hall of India Private Limited, *New Delhi*
Prentice-Hall of Japan, Inc., *Tokyo*
Simon & Schuster Asia Pte. Ltd., *Singapore*
Editora Prentice-Hall do Brasil, Ltda., *Rio de Janeiro*

Contents

Preface

Great changes in land mobile radio systems are happening as this second edition is being prepared. The first change is the FCC's proposal to refarm the spectrum. This will increase the number of channels and require new types of radio equipment. The changes are laid out well into the 21st century and will revolutionize land mobile radio systems. This edition includes a concept and strategy for dealing with these rapid changes.

Digitizing the human voice is the second important change. Second-generation cellular, simulcasting, trunking, and other new systems using digitized voice are described.

This edition takes these and other changes into account. The information on conserving spectrum in the first edition is expanded with new technologies. The FCC refarming proposals that are listed will require many of these new technologies.

This edition also expands the information on digital technologies including forward error-correcting systems. A section has been added on APCO Project 25, which will produce an open digital architecture in land mobile radio systems in the United States and many foreign countries. APCO Project 25 will use the digital technologies described in this edition.

The world is getting smaller in technological terms and land mobile radio communications will become more international in the near future. The International Telecommunication Union (ITU) laid out the future public land mobile telecommunications system (FPLMTS) concept in 1992, which is intended to become a worldwide personal communications network using satellites. The concept is described in this edition.

This modern handbook of land mobile radio systems enables the reader to understand and apply the tremendous changes integrated circuits, microprocessors, computers, digital technology, and satellites present. The book integrates the new technologies with basic information on many different land mobile radio systems, and gives solutions to prevent frustrating intermittent system failures.

The first chapter, an overall view, gives information on user organizations, frequency bands and their characteristics, applicable documents, and where to obtain them.

Conventional land mobile radio system components are described in Chapter 2. This chapter includes remote control of transmitters and what to do when the transmitter jumps frequency intermittently for no apparent reason.

Chapter 3 covers squelch systems, including a situation that can cause intermittent dropouts of syllables in the middle of a word. A solution is presented that eliminates this type of dropout the tone-coded squelch circuit causes.

Chapter 4 on duplexers and antenna combiners gives information on how they perform. Specification factors are presented that can be helpful in deciding what to buy and use.

Chapter 5's presentation of frequency control, including frequency synthesizers, can help you solve problems by using consolidated frequency control.

Chapter 6 on improving and extending area coverage describes voting systems, simulcasting, and digital computer controller wide-area systems.

The FCC is imposing spectrum conservation on the land mobile radio community because radio spectrum is becoming a scare commodity. By understanding systems that conserve spectrum, you can make better choices in the future. Chapter 7 covers trunking, cellular, amplitude, compandored single sideband, narrowband FM, narrowband digitized voice channels, spread-spectrum techniques, multiplexing, and the FCC proposals for refarming the spectrum. A concept for dealing with these rapid changes the FCC proposes is discussed.

Radio tie lines such as telephone, microwaves, fiber optics, CATV, and RF used in land mobile radio systems for voting receivers and transmitter remote control are described in Chapter 8. A method to determine when a microwave system is more economical than a telephone tie line is illustrated.

The microcomputer is revolutionizing land mobile radio systems. A review of basic computer technology is presented in Chapter 9. Some microprocessors used in land mobile radio systems are described. Computer applications in land mobile radio systems are presented, including microprocessor control of a group of consoles with a discussion of advantages and disadvantages. An example of a large computer-aided dispatch system (CAD) is described as well as a second-generation system. The entire chapter helps readers understand the changes that computers will make in their radio systems.

Digital communications is one of the technologies changing the operation of land mobile radio systems. This book presents a background in digital communications applicable to land mobile radio systems. Chapter 10 includes a survey of digital modulation, error bit detection and correction, sync versus async, channel control, digital technology applications, and packet radio technology.

Paging systems are a component of some large land mobile radio systems. In addition paging systems exist as independent business enterprises. Chapter 11 presents information including two-tone and five-tone sequential and digital systems such as NEC, POCSAG, and GOLAY.

Emergency power systems are important in keeping land mobile radio systems operating during power failures. Standby emergency electric generating systems are described, together with a maintenance schedule a large land mobile radio system uses. This maintenance schedule can help you avoid the embarrassing problem of an emergency power system that does not operate in an emergency. Also included: a description of uninterrupted power supply (UPS) and its maintenance.

Radio shop management is another important area. Chapter 13 illustrates techniques for spare parts control that can help readers manage this problem. To assist in decision-making, a method of cost analysis that can be used to compare contracting out versus in-house repairs is explained.

Using integrated circuits in land mobile radio systems and associated computers has greatly increased these systems' vulnerability to lightning. Proper grounding techniques and methods of decreasing lightning damage are described in Chapter 14.

Chapter 15 on propagation helps solve the problem of obtaining reliable coverage of an area. Computer programs for propagation are examined together with where they can be obtained. A method of measuring a system's signal quality is described.

Radio interference can degrade a radio system's operation. Examples and solutions that can help solve similar problems are given in Chapter 16.

Communications in enclosed spaces such as high-rise buildings, tunnels, subways, and shopping malls can present problems. Chapter 17 provides examples of solutions both with leaky cables and emergency (police, fire) communications without available leaky cables. New developments in enclosed-area communications are discussed.

Chapter 18 on automatic vehicle location presents information on the general principles of different methods and specific systems to assist readers in making an intelligent choice that is right for their radio systems.

Satellite communication systems in the near future will play an important part in the land mobile radio service. Chapter 19 gives important data on a new technology that has the potential of revolutionizing the field.

Finally, this book gives information on station licensing procedures, including frequency coordination in Chapter 20. This includes the definition of licensing terms, documentation, a list of frequency coordinating agencies, and engineering considerations for frequency coordination. This information can help readers obtain a station license and avoid an interference problem with another radio system.

The entire book is a source of information and problem solving that helps keep a system functioning and aids readers in facing technological change.

ACKNOWLEDGMENTS

This book would not be possible without the assistance and suggestions of my wife, Hilda Gofstein Singer.

My former colleagues in the New York City Fire Department Division of Fire Communications, have also been very encouraging and helpful.

I also wish to thank Mr. George Kormos for his assistance in preparing the figures used in this book.

I wish to thank the many companies and organizations in the field who supplied me with information, material, and advice. Among these are Andrew Corporation, The Antenna Specialists Co., Cellular Communications, Ericsson General Electric, Mobile Satellite Corp., Motorola, and the Telecommunications Industry Association.

1

Obtaining an Overview

This chapter gives a broad overview of the field of land mobile radio. It includes a list of users and users' organizations, information sources, and international organizations involved in the field. It also explains how and where to obtain standards and other documents useful in land mobile radio systems, as well as gives a list of different frequency bands and their characteristics. The final part of the overview is a survey of some of the technology that is rapidly changing the field. (This is detailed in later chapters.)

1.1 FINDING OUT WHO IS WHO

Users.   There are two general classes of users. The first includes specific groups that meet specific requirements to use land mobile radio:

> *Public safety radio services*
> Local government
> Police
> Fire
> Highway maintenance
> Forestry-conservation
>
> *Special emergency radio services*
> Medical services, including:
>   Hospitals
>   Ambulances

Rescue organizations

Clinics, public health facilities, and similar establishments providing medical services

Physicians, schools of medicine, and oral surgeons

Physically handicapped

Veterinarians

Disaster-relief organizations

School buses

Beach patrols

Industrial radio services

Power (utilities)

Petroleum

Forest products

Motion picture

Relay press, including:

Newspaper publication

Press association operation

Special industrial, including operation of

Farms, ranches

Livestock breeding services

Commercial construction of roads, bridges, pipelines, and airfields

Mines for solid fuels, minerals, etc.

Liquid transmission pipelines

Services incident to drilling new oil or gas wells, or maintaining production from established wells

Supplying materials or equipment unique to petroleum and gas production industries

Delivering ice or fuel to consumers for heating, lighting, and refrigeration

Business radio services, including operation of

Commercial activities

Educational, philanthropic, or ecclesiastical institutions

Clergy activities

Hospitals, clinics, or medical associations

Manufacturers' radio service

Telephone maintenance radio service

Land transportation radio services

Motor carrier

Railroad

Taxicab

Automobile emergency—dispatching repair trucks and tow trucks to disabled vehicles

Radio location services

Services that use radio methods for determining direction, distance, speed, or position for purposes other than navigation

Specialized mobile radio services (SMRS)

Radio systems owned and operated by the licensee. (The users are charged a fee, but they are in occupations eligible for radio station licenses.)

The second general class of users is the radio common carrier (RCC). A company can operate a system and charge a fee to use its communications facilities. The users do not have to meet any specific requirements—paging and radio mobile telephone systems, for example.

Users' Organizations.   The various users maintain associations that may represent their members before the Federal Communications Commission (FCC) and other government bodies. Many of these associations publish their own magazines or newsletters. In some cases more than one association serves a particular radio service.

Organizations

American Association of State Highway and Transportation Officials
444 North Capitol Street, NW, Suite 249
Washington, DC 20001
202-624-5800
Fax: 202-624-5469
* Covers highway maintenance radio service

American Automobile Association
National Emergency Road Services
1000 AAA Drive, Mailspace 15
City of Heathrow, FL 32746-5063
407-444-7000
Fax: 407-444-7380
* Covers automobile emergency radio service

American Hospital Association
American Society for Hospital Engineering
Telecommunications Section
840 North Lake Shore Drive
Chicago, IL 60611
312-280-5223
Fax: 312-280-6786
* Covers medical radio service

American Petroleum Institute
Information Systems
1220 L Street
Washington, DC 20005
202-682-8364
Fax: 202-682-8521
* Covers petroleum radio service

American Trucking Association, Inc.
2200 Mill Road
Alexandria, VA 22314
703-838-1730
Fax: 703-683-1934
* Covers motor carrier radio service

Associated Public Safety Communications Officers, Inc. (APCO)
2040 South Ridgewood Avenue, Suite 202
South Daytona, FL 32119
800-949-2726
Fax: 904-322-2502
* Covers public safety and special emergency radio services

Association of American Railroads
50 F Street, NW
Washington, DC 20001
202-639-2217
* Covers railroad radio service

Cellular Telecommunications Industry Association
1990 M Street, NW, Suite 610
Washington, DC 20036
202-785-0081
Fax: 202-785-0721
* Covers cellular radio systems

Forestry Industries Telecommunications
871 Country Club Road, Suite A
Eugene, OR 97401
503-485-8441
Fax: 503-485-7556
* Covers forest products radio service

Forestry Conservation Communications Association (FCCA)
444 North Capitol Street, NW, Suite 540
Washington, DC 20001
202-624-5416
Fax: 202-624-5407
* Covers forestry conservation radio service

International Municipal Signal Association (IMSA)
P. O. Box 539
Newark, NY 14513
315-331-2182
Fax: 315-331-8205
* Covers fire and special emergency radio service

International Taxicab Association
3849 Farragut Avenue
Kensington, MD 20895
301-946-5700
Fax: 301-946-4641
* Covers taxicab radio service

Land Mobile Communications Council (LMCC)
c/o Keller and Heckman
West 1001 G Street, NW, Suite 500
Washington, DC 20001
Telephone: 703-434-4100
Fax: 703-434-4646
* An association of land mobile radio user groups and equipment manufacturers

National Association of Business and Education Radio (NABER)
1501 Duke Street, Suite 200
Alexandria, VA 22314
703-739-0300
Fax: 703-836-1608
* Covers business, education, and specialized mobile radio services (SMRS)

National Association of State Telecommunications Directors (NASTD)
Ironworks Pike, P.O. Box 1190
Lexington, KY 40578
606-231-1900
Fax: 606-231-1970
* Covers statewide telecommunications

Telecommunications Industry Association (TIA)
2001 Pennsylvania Avenue, NW, Suite 800
Washington, DC 20006
202-457-4912
Fax: 202-457-4939
* Represents manufacturers of telecommunications equipment. TIA together with the
    Electronic Industries Association (EIA) prepares standards for the telecommuni-
    cations industry.

Telocator Network of America
1019 19th Street, NW, Suite 1100
Washington, DC 20036
202-467-4770
Fax: 202-467-6987
* Covers radio common carrier services

Utilities Telecommunications Council (UTC)
1620 I Street, NW, Suite 515
Washington, DC 20006
202-872-0030
Fax: 202-872-1331
* Covers power radio service

Information sources.   In addition to specific user organizations there are other
sources of information, including international organizations that issue frequency blocks
and general radio regulations that become international treaties when ratified by the
affected governments.

IEEE Vehicular Technology Society
c/o IEEE Headquarters
345 East 47th Street
New York, NY 10017

National Telecommunications Information Administration (NTIA)
U.S. Department of Commerce
14th Street and Constitution Avenue, NW
Washington, DC 20230
* This is the United States government's telecommunications policy office. It also assigns and administers frequencies to federal agencies.

The Radio Club of America
P. O. Box 4075
Overland Park, KS 66204-0075

International organizations. These three deal with telecommunications matters:

The *International Telecommunications Union (ITU)* is a treaty organization affiliated with the United Nations. Its charter includes the regulation and use of the radio spectrum. Its policies and procedures are established through a Plenipotentiary Conference, which in turn delegates much of its responsibility to a 29-member administrative council that meets annually and acts for and on behalf of the Plenipotentiary Conference. World Administrative Radio Conferences for all members are called on special issues as they arise.

The *International Radio Consultative Committee (CCIR)* is one of the ITU's permanent organs. The CCIR acts primarily with to improve radio communications by establishing technical specifications. It works principally through permanently established study groups comprised mostly of volunteer experts from many countries.

The *International Telegraph and Telephone Consultative Committee (CCITT)* is similar to the CCIR but deals with telephone and telegraph communications.

1.2 CHARACTERISTICS OF DIFFERENT FREQUENCY BANDS

VHF low-band. From 25 to 50 MHz the bandwidth is generally 20 kHz.

VHF high-band. From 72 to 76 kHz the bandwidth is 30 kHz. From 150 to 174 MHz the bandwidth is 30 kHz, also, but with the use of geographical separation, the adjacent carrier frequencies are only 15 kHz apart.

UHF. This is from 450 to 470 MHz, and bandwidth is 25 kHz. Frequency separation between mobile and base is 5 MHz.

UHF-T. This is from 470 to 512 MHz and is shared with UHF-TV stations 14 through 20 in a few major urban areas. Bandwidth is 25 kHz. The details are in the FCC Rules, Part 90, subpart L. Here are the pertinent UHF-TV channels for reference:

UHF-TV channel	Frequency (MHz)
14	470–476
15	476–482
16	482–488
17	488–494
18	494–500
19	500–506
20	506–512

800/900-MHz band.   This band is the latest authorized one (except for the mobile satellite service). If systems have more than five channels, they must use trunking. The base transmitter frequencies are 45 MHz above the corresponding mobile transmitter frequencies. An exception: the private radio mobile service (nonpublic safety), 896 to 901 MHz, which is 39 MHz above the corresponding private radio base transmitters.

Since there is great interest in the 800/900-MHz band, frequency assignments are shown below. These are subject to revision by the FCC.

Frequency assignments	800/900-MHz band
806–809.750	Conventional mobile
809.750–816	Conventional and trunked mobile
816–821	Trunked mobile
821–824	Public safety mobile
824–825	Cellular wire line mobile
825–835	Cellular nonwire line mobile
835–845	Cellular wire line mobile
845–846.5	Cellular nonwire line mobile
846.5–849	Cellular wire line mobile
849–851	Reserved
851–854.750	Conventional base
854.750–861	Conventional and trunked base
861–866	Trunked base
866–869	Public safety base
869–870	Cellular wire line base
870–880	Cellular nonwire line base

Frequency assignments	800/900-MHz band
880–890	Cellular wire line base
890–891.5	Cellular nonwire line base
891.5–894	Cellular wire line base
894–896	Reserved
896–901	Private radio mobile (nonpublic safety)*
901–902	New general-purpose mobile radio service
902–929	Industrial—scientific—medical
929–930	Private radio paging
930–931	New technology paging
931–935	Common carrier paging
935–940	Private radio base (nonpublic safety)*
940–941	New general-purpose base radio service

* Includes specialized mobile radio services 5 MHz, business 2.5 MHz, and industrial/land transportation 2.5 MHz.

National public safety planning advisory committee (NPSPAC) channels. This nationwide public safety plan has been implemented in the United States for the 821–824 MHz and 866–869 MHz bands. Mobile and control stations use 821–824 MHz while talk-around and base stations use 866–869 MHz.

The plan separates base stations geographically to obtain more channels. The center frequencies of two adjacent base transmitters are 25 kHz apart. A third base transmitter is separated geographically from the other two by 40 miles, and the center frequency of this third base transmitter is 12.5 kHz from the center frequencies of the others. The term used in public safety communications is *public safety answering point (PSAP)*.

Brief Comparison of Bands

The frequency bands differ in noise levels, ranges, skip, and other factors.

VHF low-band. This band is subject to heavy skip—signals bounce off the ionosphere and travel great distances. There are frequent dead spots, and the signal does not bounce off hills or buildings. It has the most range and the highest noise level.

VHF high-band. There is much less skip, less range, and less noise in this band than VHF low band. This band has fewer dead spots than the VHF low-band, too.

UHF band.   This band's range is less than in VHF high-band. The signal bounces off hills and buildings well and has practically no skip interference. It too has fewer dead spots and less noise.

800/900-MHz band.   The signal bounces off buildings and hills extremely well and presents little noise. The range is even smaller than UHF's and there is more absorption by foliage.

1.3 REGULATIONS AND STANDARDS

FCC Rules and Regulations in Land Mobile Radio Services

The following FCC rules and regulations, which are part of the Code of Federal Regulations, Telecommunication 47, are important when licensing and operating land mobile radio systems.

Part 22, Public Mobile Radio Services, spells out the radio common carriers in which a company can operate a system and charge users a fee to use its communication facilities. The users do not have to meet any specific requirements.

Part 90, Private Land Mobile Radio Services, includes various licensed user services, such as taxi, fire, police, and so on. Specialized mobile radios (SMRS) owned by an entrepreneur are included also. The entrepreneur rents the services of the radio system to users listed in Part 90. Only the SMR is licensed. The FCC soon will replace Part 90 with a new Part 88.

Applicable Standards for Land Mobile Radio Systems

These standards are from EIA, TIA, and IEEE. You may obtain them from Global Engineering Documents, 15 Inverness Way East, Englewood, Colorado 80112-5704, telephone 800-624-3974, fax 303-267-1326.

The following lists some of the standards:

Standard number	Standard title
EIA 152-C	Minimum Standards for Land Mobile Communication, FM or PM Transmitters
EIA 195-C	Electrical and Mechanical Characteristics for Terrestrial Relay System Antennas and Passive Reflectors
EIA/TIA 204-D	Minimum Standards for Land Mobile Communications, FM or PM Receivers, 25 to 866 MHz
EIA 220B	Minimum Standards for Land Mobile Communications Continuous Tone-Controlled Squelch Systems (CTCSS). (This revision also accommodates Equipment Operating Within a Maximum Frequency Deviation of 2.5 kHz.)

Standard number	Standard title
EIA/TIA 232-E	Interface Between Data Terminal Equipment and Data Circuit Terminating Equipment Employing Serial Binary Data Interchange (formerly RS 232).
EIA 252-A	Standard Microwave Transmission Systems
EIA 258	Semi-Flexible Air Dielectric Coaxial Cables and Connectors, 50 Ohms
EIA/TIA 316-C	Minimum Standards for Portable/Personal Radio Transmitters, Receivers, and Transmitter/Receiver Combination Land Mobile Communications FM or PM Equipment, 25 to 1000 MHz
EIA/TIA 329-B	Minimum Standards for Communication Antennas, Part 1 - Base Station Antennae
EIA/TIA 329-B-1	Minimum Standards for Communication Antennas, Part II - Vehicular Antennas
EIA 368	Frequency Division Multiplex Equipment Standard for Nominal 4 kHz Channel Bandwidths (Non-Compandored) and Wideband Channels (Greater Than 4 kHz)
EIA/TIA 374-A	Land Mobile Signalling Standard
EIA 411-A	Electrical and Mechanical Characteristics of Earth Station Antennas for Satellite Communications
EIA 440-A	Fiber Optic Terminology
EIA 450	Standard Form for Reporting Measurements of Land Mobile, Base Station, and Portable/Personal Radio Receivers in Compliance with FCC Part 15 Rules
EIA IS 52	Uniform Dialing Procedures and Call Processing Treatment for Use in Cellular Radio Telecommunications
EIA/TIA IS 53	Cellular Features Description
EIA/TIA IS 54-B	Cellular System Dual-Mode Mobile Station-Base Station Compatibility Standard. (This includes TDMA Digital Cellular Technology.)
EIA/TIA IS 55	Recommended Minimum Performance Standards for 800 MHz Dual-Mode Mobile Stations
EIA/TIA IS 56	Recommended Minimum Performance Standards for 800 MHz Base Stations Supporting Dual-Mode Mobile Stations
ANSI/IEEE C95.1-1992[*]	Safety Levels with Respect to Human Exposure to Radio Frequency Electromagnetic Fields, 3 kHz to 300 GHz

* The FCC on April 8, 1993 proposed adopting this standard.

Publications by the National Institute of Justice, United States Department of Justice

Many of these publications cover components and techniques in police radio communications, but most are applicable in general to land mobile radio systems. A catalog, *Putting*

Research to Work, is available from the Superintendent of Documents, U.S. Government Printing Office, Washington, DC 20402.

1.4 GETTING ACQUAINTED WITH NEW TECHNOLOGIES

History

On April 7, 1928, the city of Detroit, Michigan, began operating a one-way radio system for its police cars. Two-way police communications began in Bayonne, New Jersey, five years later. In 1940, the Connecticut State Police Department at Hartford began operating an improved two-way FM system, which stood as the model for many years.

Solid-state circuitry began to replace vacuum tubes in the 1960s, and in 1969 the market introduced tone-coded squelch. But by the middle 1970s the introduction of very large-scale integrated (VLSI) circuit chips—with more than 1,000 logic gates, such as AND or OR circuits—changed the face of telecommunications. It was the development of the VLSI chip that facilitated the microprocessor.

Development of microprocessors. The first commercial microprocessor, the Intel 404, a 4-bit unit, came out in 1971. This was followed by the 8-bit Intel 8008. Improvements in chip technology then increased the instruction execution speed by a factor of five, prompting the Motorola 6800 and the Intel 8080 8-bit units. The advanced 8-bit chip Zilog Z80 was introduced in 1976. All three of these 8-bit microprocessors have been used in land mobile radio systems.

In 1978, 16-bit microprocessors were introduced, followed by the commercially produced Motorola 68000 in 1980. In 1984, Motorola introduced a 32-bit unit, the 68020. Zilog followed with its 32-bit microprocessor, the Z80000 that same year. Almost all of these 8-bit, 16-bit, and 32-bit microprocessors have been used with other VLSI chips to develop new technologies in land mobile radio systems.

Present-day Developments

Spectrum conservation. The demand for spectrum has led to techniques such as cellular and amplitude-compandored single sideband (ACSB). Both cellular and trunking depend on the computer, while ACSB is made possible by VLSI chips. Multiplexing techniques such as frequency division multiple access (FDMA), time division multiple access (TDMA), and code division multiple access (CDMA) now are used in land mobile radio systems to conserve spectrum, also. Narrowband digitized voice systems have also been developed to decrease bandwidth to 12.5 kHz and may in the future decrease bandwidth to 6.25 kHz. Advanced digital signal processing (DSP) has made narrowband digitized voice practical.

Automatic vehicle location (AVL). Several AVL techniques such as Loran C, Dead Reckoning, and the Global Positioning System (GPS) use microprocessors and VLSI to send the location of a vehicle to a dispatching center automatically. A computer can combine this information with computer-aided dispatching to obtain an efficient dispatching center.

Frequency synthesizers.   The frequency synthesizer enables small walkie-talkies to use many frequencies, thanks to microprocessors and VLSI chips.

Computer-aided dispatching (CAD).   In computer-aided dispatching the computer maintains information on the location of mobile units and recommends dispatching of specific units. Microprocessors as well as large computers have made CAD practical in land mobile radio systems.

Console control.   Microprocessors now are used in many large system consoles to control radio frequencies, coded squelch tones, transmitters' selection, voting receivers, and so on.

Digital communications.   The microprocessor has made possible digital communications in land mobile radio systems. This includes mobile data terminals (MDTs), which allow transmission as well as receive digital communications. A system of radio communications from a portable computer unit to a central computer is included, too. Packet technology has been developed for use in land mobile radio systems. Using computers and other VLSI chips, you can convert a conventional FM voice radio system to packet digital communications.

Mobile satellite service (MSS).   This new service for land mobile radio systems uses satellites in a geostationary orbit to communicate with mobile units. In addition, the satellites can locate the positions of mobile units and relay them to a dispatching center. All of this activity is controlled by computers.

REFERENCES

Duff, William G. *A Handbook on Mobile Communications.* Germantown, MD: Don White Consultants, Inc., 1980.

Jakes, William C., Jr. *Microwave Mobile Communications.* New York: John Wiley & Sons, Inc., 1974.

Lee, William C.Y. *Mobile Communications Design Fundamentals,* 2d Edition, New York: John Wiley & Sons, 1992.

_____ . *Mobile Communications Engineering.* New York: McGraw-Hill Book Company, 1982.

Pannell, William M. *Frequency Engineering in Mobile Radio Bands.* Cambridge: Granta Technical Editions, 1979. Available in the United States from Telecom Library Inc., 205 West 19 Street, New York, NY 10011.

2

Examining Conventional FM Mobile Radio Systems

This chapter covers simplex, duplex, and semiduplex operations. It describes system components, including transmitters, receivers, mobile relays, portables, antennas, transmission lines, and towers.

Specification definitions and typical values are given to help in both selecting equipment and understanding its operation.

Guidelines are presented for mounting antennas on a tower. These include the vertical separation of antennas on the tower for each frequency band.

Consoles and associated equipment are described, with definitions of console terminology.

Different methods of remotely controlling transmitters are discussed. A frequency-jumping problem in a remote-controlled transmitter is described, along with a method for avoiding this frustrating problem that has plagued many land mobile radio systems.

2.1 SIMPLEX AND DUPLEX SYSTEMS

In radio communications these terms[*] do not have exactly the same definition as in other forms of communications.

Simplex Operation

In simplex operation one terminal of the system transmits while the other terminal receives. Simultaneous transmission and reception at a terminal is not possible with simplex operation.

_____

* Simplex and duplex systems here are based on the definitions in 2.1 of Part 2 of the FCC Rules, revised as of October 1985. Two-frequency simplex, as defined here, is sometimes referred to as half duplex. Duplex, as defined here, is sometimes referred to as full duplex.

The single-frequency simplex dispatching system consists of a base station and mobile units, all operating on a single frequency.

In the two-frequency simplex dispatching system, the base station operates on a different frequency from the mobile unit. However, the terminal equipment does not allow simultaneous transmission and reception. This may be because the antenna is switched from the receiver to the transmitter when transmitting or because some other operating equipment at the terminals prevents simultaneous transmission and reception.

The push-to-talk switch (PTT) on a microphone is an example of terminal operating equipment that prevents simultaneous transmitting and receiving. When pushed, the PTT transfers direct current (dc) voltages from the receiver to the transmitter and also transfers the antenna from the receiver to the transmitter.

Duplex Operation

In duplex operations, terminals have the ability to receive while transmitting. This requires two frequencies and equipment such as a duplexer, which enables the receiver and transmitter to use the same antenna simultaneously.

Semiduplex dispatching system. In a semiduplex dispatching system, the terminal equipment at the base station permits duplex operation, but the terminal equipment at the mobile permits only simplex operation. For example, the base station has a duplexer that permits simultaneous reception and transmission, while the mobile uses PTT.

2.2 TRANSMITTERS

Most transmitters in land mobile radio systems use phase modulation to produce frequency modulation. However, there are transmitters that use direct frequency modulation.

Understanding the difference between phase and frequency modulation helps when applying new technologies. This is the case when conventional voice transmitters are modified for digital operation.

Phase Versus Frequency Modulation

Frequency modulation can be produced directly by the audio signal varying the oscillator's radio frequency or produced indirectly by phase modulation. In phase modulation the audio varies the radio-frequency (RF) carrier's phase. In both frequency and phase modulation the frequency deviation of the RF carrier varies directly with the modulating audio signal's amplitude. In phase modulation, however, the RF carrier's frequency deviation also varies directly with the modulating audio signal's frequency. This results in a preemphasis of the higher audio frequencies, which is used to increase the signal-to-noise ratio. This occurs because noise encountered in signal transmission is mainly at the higher audio frequencies. A deemphasis circuit in the receiver compensates for the preemphasis. Phase modulation produces a very small frequency deviation that must be multiplied to usable levels. Phase modulation is, therefore, often used with low-cost, low-frequency RF crystals followed by a number of frequency-multiplying stages that also multiply the frequency deviation. Figure 2-1 illustrates phase modulation, which is used in most base transmitters. A low-frequency crystal-controlled oscillator is phase modulated by an audio signal. The amount of the RF carrier's frequency deviation is small. The frequency multipliers increase both the carrier frequency and frequency deviation to the desired levels. A driver stage supplies the input to a radio-frequency power amplifier. FCC regulations require the harmonic filter.

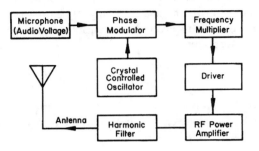

Fig. 2-1 Phase-modulated Transmitter

Figure 2-2 illustrates direct production of FM used in some mobiles. The audio voltage modulates a voltage-controlled oscillator (VCO) to produce an FM signal directly. In this particular case a frequency synthesizer is used to select a number of frequencies.

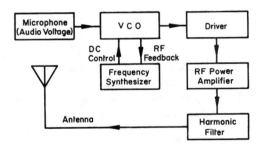

Fig. 2-2 Transmitter with Direct Frequency Modulation

Transmitter Performance Specifications

It is necessary to understand performance specification terms when selecting transmitters as well as other equipment:

Spurious and harmonics. Spurious emission is any part of the RF output that is outside the assigned frequency band. A typical value is 85 decibels[*] (dB) below the carrier.

---

* The decibel is explained in Appendix I.

FM noise.   FM noise is the frequency modulation present on an unmodulated carrier. A typical value is 55 dB below 60% system deviation at 1000 Hz.

Transmitter sideband noise.   Transmitter sideband noise is measured at different separations from the carrier without audio modulation. Typical values at VHF high-band are -90 dB at ±30 kHz and -105 dB at ±1 MHz.

Audio response.   The frequency response for audio is from 300 to 3000 Hz. Pre-emphasis increases the higher audio frequencies relative to the lower frequencies. The pre-emphasis is 6 dB per octave. This means that doubling the audio frequency doubles the audio voltage. A typical specification is +1, -3 dB tolerance for a 6 dB per octave response.

Audio distortion.   Audio distortion is a measure of the audio harmonics. A typical value is 3% at 1000 Hz, at 60% system deviation.

Frequency deviation.   The frequency deviation is the peak difference between the instantaneous frequency of the modulated wave and the carrier frequency in frequency modulation. The maximum amount for FM voice transmitters is ±5 kHz.

RF output power.   The RF output power is the power available at the antenna terminal with no modulation. FCC regulations determine the maximum allowed power.

Oscillator frequency stability.   Oscillator frequency stability is the frequency tolerance within a specific temperature range. The allowable frequency tolerance[*] depends on the band and power output. A typical specification for VHF high-band is ±0.0005%, with an option of ±0.0002%, from -30°C to +60°C ambient.

Frequency separation.   Frequency separation is the maximum separation between transmitting frequencies that can operate without degradation. A typical value is 10 MHz, but it will vary with equipment. The value for transmitter frequency separation is not the same as the value for receiver frequency separation.

2.3 RECEIVERS

A typical single-conversation receiver is shown in figure 2-3. The receiver is a superheterodyne in which the incoming radio-frequency signal is mixed with a local oscillator frequency in a mixer circuit. The mixer's usable output is the difference frequency, which is called the intermediate frequency (IF). The IF is amplified in an amplifier that is tuned to the IF and then sent to a detector, where the audio signal is extracted and amplified. The IF amplifier may also be untuned, but then it is preceded and followed by IF filters. The original radio-frequency signal is converted to a fixed IF, which is easier to amplify. The IF in

---

[*] FCC frequency tolerances are listed in Chapter 5.

Sec. 2.3 Receivers

this case is fixed at 10.7 MHz no matter what the RF preamp and mixer are tuned to. In this example, the local oscillator always is tuned to 10.7 MHz about the incoming RF signal. This is a single-conversion receiver. Frequency selection is at the local oscillator stage.

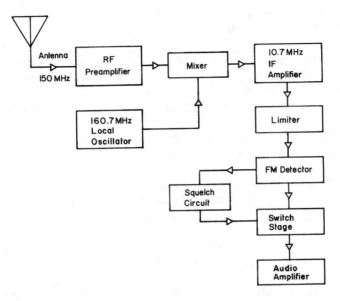

Fig. 2-3 Typical Conventional FM Receiver

Most of the receivers now on the market are single-conversion types. In this case the RF signal is converted to an IF signal as shown in figure 2-3. Another local oscillator frequency is then mixed with the IF signal to produce a second IF, which is amplified and detected as before. Frequency selection is at the first local oscillator frequency.

The squelch circuit indicated in figure 2-3 may be one of three types: carrier squelch, tone-coded squelch, or digital squelch. In carrier squelch the receiver noise is converted to a dc voltage that cuts off the audio in the absence of an RF carrier. When an RF carrier is present, the RF noise is quieted and the resultant lower dc voltage permits the audio to be activated. In tone-coded squelch, the audio circuit is activated only when certain subaudible tones are received. Tone-coded squelch prevents listeners from hearing undesired RF signals. Digital squelch does the same by using digital codes. The details of squelch circuits, including specific problems, are discussed in Chapter 3.

Receiver Performance Specifications

These specifications are important in equipment selection:

Receiver sensitivity. Two methods are used to measure receiver sensitivity: 20-dB quieting and EIA 12-dB SINAD sensitivity. The first, 20-dB quieting, is the amount of the radio-frequency signal that reduces noise at the audio output by 20 dB. A typical value is 0.5 microvolt (μV). SINAD is the ratio in decibels at the receiver audio output, of the signal plus noise plus distortion to noise and distortion. A SINAD of 12 dB represents a fair signal, while a 20-dB SINAD is a good signal. A typical specification is 0.35 μV for EIA 12-dB SINAD.

Channel spacing. Receiver channel spacing is 20 kHz for VHF low-band, 30 kHz for VHF high-band, and 25 kHz for UHF and 800 to 900 MHz.

Modulation acceptance bandwidth. The modulation acceptance bandwidth is the deviation the receiver passes with an RF signal 6 dB above the receiver's usable sensitivity. The usable sensitivity is the minimum RF input signal that produces at least 50% of the rated audio power output of the receiver with a 12-dB SINAD. This is from EIA RS-316. A typical value of modulation acceptance bandwidth is ±7 kHz.

Adjacent channel selectivity and desensitization. Adjacent channel selectivity is a receiver's ability to discriminate between a desired signal and an undesired signal in an adjacent channel. Desensitization happens when a very large signal in an adjacent channel is so great, it reduces the receiver's sensitivity by overloading it, changing biases, and so on. A typical specification value for adjacent selectivity and desensitization is -95 dB.

Intermodulation attenuation. Intermodulation attenuation is the ratio in decibels of the RF signal that produces intermodulation to the receiver's 12-dB SINAD sensitivity. A typical value for intermodulation attenuation is -80 dB.

Spurious response attenuation. Spurious response is the output of a receiver caused by a signal at a frequency other than the one the receiver is tuned to. A typical specification value is an attenuation of 100 dB.

Oscillator frequency stability. Oscillator frequency stability is the same as for transmitters in Section 2.2.

Frequency separation. Frequency separation is the maximum separation between receiver frequencies that can operate without degradation. A typical value is 2 MHz.

Audio characteristics. These include:

Output for telephone line (base or remote receiver). A typical value is +18 dBm at 600 ohms (Ω).

Audio output and distortion at speaker. A typical audio output for a base station receiver is 5 W at 16 Ω with 5% distortion. For a portable receiver a typical output is 500 mW with 5% distortion.

Audio hum and noise. A typical value is -50 dB.

Audio response. A typical value is +2, -8 dB of a 6-dB per octave deemphasis response from 300 to 3000 Hz.

Environmental requirements. The receiver (and transmitter) should be capable of operating within specifications from -30 to +60°C. The equipment should also be capable of operating with a humidity of 90% at 50°C.

2.4 MOBILE RELAY

Figure 2-4 shows a block diagram of a mobile relay that serves as a fixed repeater for mobile units. When the repeater receives a signal from mobile 1, it activates a squelch relay. The squelch relay turns on the repeater transmitter, which forwards the receiver audio to mobile 2.

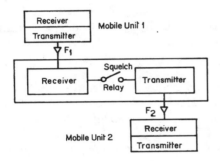

Fig. 2-4 Mobile Relay

Although not shown in the diagram, mobile 2 also can transmit to mobile 1 through the mobile relay. F_1 is the frequency of all the mobile transmitters and the mobile relay receiver. F_2 is the frequency of all the mobile receivers and the mobile relay transmitter.

Repeaters should not be used without tone-coded squelch or digital-coded squelch to prevent extraneous signals inadvertently turning on the repeater. The squelch relay in figure 2-4 is actually in the receiver. The relay is sometimes known as carrier-operated relay (COR).

The mobile relay may be used as a base station operating with a dispatcher. Usually the equipment gives priority to the dispatching function.

2.5 PORTABLES

Hand-held portable radios—also known as walkie-talkies—are used in many mobile radio systems. Portables are used with the mobile relay and in direct portable-to-portable communications. In a two-channel walkie-talkie, channel 1 may be used for repeater operation, called "talk thru." Channel 2 may be used to talk directly from one portable to another, called "talk around." The frequencies are

Channel 1: transmitter, F_1; receiver, F_2
Channel 2: transmitter, F_3; receiver, F_3

Portable Radio Batteries

Most portable two-way radios use the nickel-cadmium chargeable battery. Proper handling of these batteries, as described later in the section "Special Devices to Cure Memory Effects," can save money and increase the efficiency of operation. The battery consists of a number of cells, which typically range from 6 to 12, depending on the battery's voltage. Nickel-cadmium batteries have a standard 1.25 V per cell in a fully charged condition. In all cases the battery never should be discharged below 1 V per cell—this results in irreparable damage to the battery.

In its basic form, the battery consists of an uncharged positive plate containing nickelous hydroxide and an uncharged negative plate containing cadmium hydroxide. The ion-conducting electrolyte is potassium hydroxide. When charged, the positive plate becomes nickelic hydroxide and the negative plate becomes cadmium. The plates are constructed by sintering a finely divided nickel powder to produce a honeycomb structure with many open pores. These pores then are impregnated with the active material so that there is enough space for this electrolyte to penetrate.

Many nickel-cadmium batteries for two-way radios distinguish separate terminals for charging from the terminals of the battery that connect directly to the radio. For example, a battery may have two metallic terminals that connect directly to the radio and four different metallic contacts for charging. Inside the battery a thermistor or a thermal fuse may be connected to the charging terminals to prevent excessive current to the battery. There is also a diode that keeps exposed charging terminals from discharging. In some radios where the battery-charging terminals are exposed when physically connected to the radio, this diode may be important in preventing injury. For example, a law enforcement officer had put such a radio in a pocket that contained bullets. A different manufacturer's replacement battery did not have the diode in it. The battery current flowed through a bullet, causing it to explode and injure the officer. Some important items to consider in using nickel-cadmium batteries are battery capacity, operation duty cycle, and memory effect:

Portable battery capacity. Battery life between charges is often given in terms of hours based on a percent transmit, percent receive, and percent standby. For example, for a 10-10-80 (10% transmit, 10% receive, 80% standby), a manufacturer may rate a battery as seven hours. This means the battery produces sufficient voltage to operate both transmitter and receiver for seven hours using the given duty cycle. The capacity of a nickel-cadmium battery is given also in milliampere-hours (mAh). Typical rated values are 450 mAh for a 15-V (12-cell) battery and 800 mAh for a 7.5-V (6-cell) battery. A battery's capacity after use varies considerably from the rated value due to memory effects.

Memory effect. A nickel-cadmium battery exhibits a memory effect when repetitive shallow cycling (discharge-charge) reduces its apparent capacity. If the battery is called on repeatedly to deliver 30% of its capacity and then charged, the remaining 70% temporarily can become inactive. The battery may then show a sharp decrease in terminal voltage when called on to deliver. The cure for "memorized" batteries is a series of full discharge-charge cycles. The full discharge is down to 1V per cell.

Special devices to cure memory effects. This equipment automatically cycles, first discharging the battery down to the limit of 1 V per cell and then charging the battery fully. A meter reads out the battery's capacity in percent after the discharge part of the cycle. A new battery should have at least 80% of its rated capacity. A battery in use should read at least 75%. If it is less than 75%, the battery should be cycled again. Three cycles may be necessary to cure the memory effect and bring up the capacity. The discharge-charge cycle can be used to screen out bad batteries, also.

Time-Out Timers for Portables

The "stuck button" problem occurs when the transmitter cannot be turned off by releasing the push-to-talk button. This disrupts communications and causes havoc in public safety activities such as fire and police.

Walkie-talkie manufacturers offer a time-out timer option. The time-out timer turns off the transmitter after a specific time interval. An audio tone alerts the portable operator that the transmitter is disabled, perhaps because the operator spoke beyond the time-out timer interval. The operator then releases the PTT button and the audio tone should disappear. After a brief pause, the operator can push the PTT button again and resume communicating. If the audio tone remains on after the operator releases the PTT button, the radio needs to be repaired. Experience has shown that a time interval of 30 seconds is sufficient in operations for communications with portables. Vehicular radios also should be equipped with time-out timers using an interval of 1 minute.

2.6 ANTENNAS

An antenna takes radio energy from a transmitter via an antenna transmission line and radiates the radio energy into space. The antenna also receives radio energy and feeds it down the transmission line to the radio receiver. Antennas are reciprocal: they receive and transmit equally.

Figure 2-5 shows the basic vertical dipole antenna. A half wavelength is calculated as follows:

$$\text{length (ft)} = \frac{492}{\text{frequency (MHz)}}$$

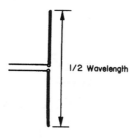

Fig. 2-5 Basic Vertical Dipole

The length of the half wave and quarter wave antennas in practice are reduced by a small percentage (in the range of 5 to 10%) of theoretical values. The reduction is due to three factors that shorten the length of the antennas. The first is the ratio of the length to the antenna's diameter. As the antenna's diameter increases, the length of the antenna must be shortened from the theoretical value. The second factor is the end effect that results from the boundary condition at the end of the radiator length where the metal ends and the air begins. At this point there is a concentration of electric lines of force that causes a greater capacitance at the end of the antenna. This reduces the antenna's physical length from the theoretical length. The third is the effect of a counterpoise or ground plane. The larger the counterpoise, the more the antenna's length is reduced from the theoretical value.

Antenna Characteristics

Antennas exhibit polarization, antenna patterns, gain, and impedance:

Polarization. Polarization is the orientation of the electric field. When the electric field of a dipole antenna is vertical to Earth, the antenna is vertically polarized. If the electric field is horizontal to Earth, it is horizontally polarized. This is illustrated in figure 2-6. Most antennas in mobile communication systems are vertically polarized.

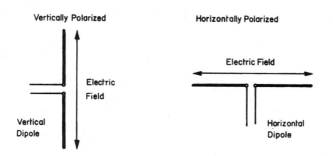

Fig. 2-6 Vertical and Horizontal Polarization

Antenna patterns. The actual antenna radiation is a solid figure. However, it is usually represented by polar graph plots only in horizontal and vertical planes. The units of the graph are angles in degrees versus relative field strength. Figure 2-7 shows the vertical radiation pattern of a vertical dipole. The horizontal pattern is a circle and is omnidirectional.

Sec. 2.6 Antennas

The vertical pattern is a figure 8 on its side with no radiation up or down. The three-dimensional picture is a horizontal doughnut with a small hole.

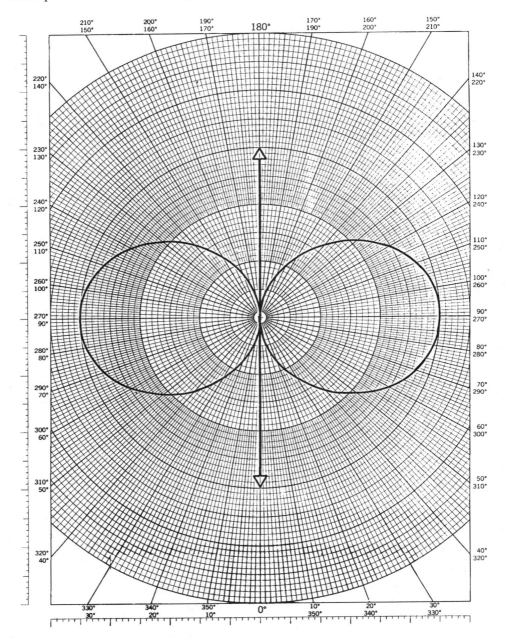

Fig. 2-7 Vertical Dipole, Vertical Pattern (Arrows Show Direction of Antenna)

Gain of an antenna.   There are two reference antennas that are used with the gain of an antenna. One is the dipole antenna and the other is a theoretical isotropic antenna with equal radiation in all directions in all planes. Antenna power gain is expressed in dB above a reference antenna. The dipole itself is 2.1 dB above the theoretical isotropic antenna. In mobile radio systems the dipole antenna is usually the reference antenna. The general way to obtain gain is to stack one vertical antenna on top of another with the proper phase shift, forming a collinear array. The vertical antenna pattern is narrowed, which results in more gain. The horizontal pattern remains a circle. For example, a dipole has a vertical beamwidth[*] of 78°. A collinear antenna with a gain of 10 dB has a corresponding vertical beamwidth of only 5.5°. The horizontal pattern is still a circle.

Antenna impedance.   Mobile radio system antennas have an impedance of 50 Ω.

Mobile Antennas

Many vehicular antennas are quarter-wave metal rods or whips mounted on a metal roof. The metal roof of the vehicle acts as a ground plane, causing a mirror image. Thus the antenna behaves as a half-wave dipole with an omnidirectional pattern in the horizontal plane. If the quarter-wave antenna is mounted on some other part of the vehicle, such as a fender, bumper, or the rear, the horizontal pattern is distorted. The antenna's gain is then greater in one direction than another.

The optimum mounting position[†] is the center of the metal roof of a vehicle. However, this is not feasible for low-band VHF antennas, because of their height. Mobile antennas vary with the frequency band.

Low-band VHF (25 to 50 MHz).   Low-band VHF uses two basic types of mobile antennas: the quarter-wave whip antenna—at 30 MHz this is about 8 ft. long—and the base-loaded antenna. A coil at the base of this antenna adds electrical length permitting a shorter physical length. However, shorter length results in a less efficient radiator than a full quarter-wave whip. Figure 2-8 shows a typical base-loaded antenna. Instead of being 8 ft long, the antenna is about 4 ft long because of base loading.

High-band VHF (130 to 174 MHz).   Many of the vehicular antennas in the high-band VHF are a quarter-wave long. At 150 MHz this is about 1 1/2 ft. There are half-wave and also 5/8-wave-long antennas with a gain of 3 dB. These 3-dB antennas have a coil at the base that acts as a matching impedance to the receiver antenna output.

UHF (450 to 512 MHz).   The length of the quarter-wave whip at 450 MHz is approximately 6 inches. Figure 2-9 shows a gain antenna that is used at UHF. Two sections are stacked in a collinear fashion with a phasing coil to obtain the proper current distribution.

---

[*] Beamwidth is measured at half-power points.

[†] See NIJ Report 201-85 in References for a study of the subject.

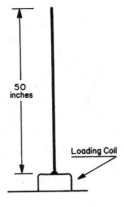

50 inches

Loading Coil

Fig. 2-8 Low-Band VHF Base-Loaded Mobile Antenna

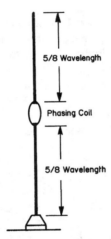

5/8 Wavelength

Phasing Coil

5/8 Wavelength

Fig. 2-9 Collinear UHF Mobile Antenna, 5 dB Gain

800/900-MHz band. The quarter-wave antenna does not perform as well in the 800/900-MHz band as it does in the VHF and UHF bands. Figure 2-10 shows a 3-dB gain collinear antenna used in the 800/900-MHz band. The upper section is 5/8 of a wavelength and the lower section is 1/4 of a wavelength. A phasing coil is used between the two sections to maintain the proper current distribution.

Base-Station Antennas

There are two general types of base-station antennas: the unity-gain antenna and one with a gain of more than 1.

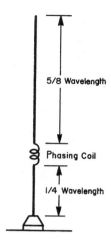

5/8 Wavelength

Phasing Coil

1/4 Wavelength

Fig. 2-10 Collinear 800-MHz Mobile
Antenna, 3 dB Gain

Unity-gain antennas. Two of the most popular unity-gain base-station antennas are the quarter-wave ground-plane antenna and the sleeve antenna, shown in figure 2-11. The quarter-wave ground-plane antenna uses quarter-wavelength radials to provide a ground plane. The antenna behaves as a half-wave dipole with a circular horizontal pattern and a unity gain. The coaxial antenna is a center-fed antenna that also behaves as a vertical dipole. The angle of the ground-plane rods, measured downward from the horizontal, determines the impedance of the feed point.

Base station gain antennas. The coaxial collinear antenna is a very popular base-station gain antenna. It is comprised of a series of solid dielectric coaxial sections with inner and outer conductors transposed at each junction. Each section has an effective length of 1/2 wavelength. Figure 2-12 shows the basic idea, but it is not intended as an actual antenna. The "outside" notation refers to the length of a half-wave in air, and "inside" refers to the length of a half-wave in the dielectric. The actual radiation is from the outside of the coaxial sections. The more stacked sections, the greater the gain but the narrower the vertical beamwidth. The current distribution on the outside of the antenna approximates that of dipole antennas stacked one above the other and all in phase.

Base Station Downward Tilt Antennas

VHF and UHF base-station antennas are mounted in high locations to maintain line-of-sight transmission. However, in vertically polarized gain antennas the beam in the vertical plane is compressed to obtain gain. The more gain, the more compression and the more minor lobes with pattern nulls in the vertical plane. This can result in dead spots at distances close to a base-station antenna located in a high place. One method for correcting this is to use a small (2 to 3 degree) amount of downward tilt. This not only eliminates the close-in dead spots but makes more use of available power and reduces interference with systems beyond the coverage area. Use antenna catalogs to select base-station antennas with downward tilt when the situation warrants it.

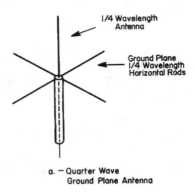

a. – Quarter Wave
Ground Plane Antenna

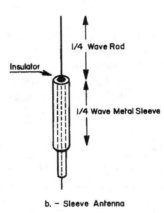

b. – Sleeve Antenna

Fig. 2-11 Base-Station Unity-Gain
Antennas

Base-Station Antenna Fiberglass Radomes

The radiating elements of many base-station antennas are enclosed in fiberglass radomes that protect the antenna from the elements. They also minimize precipitation static caused by particles of rain or snow striking an exposed antenna and creating interference, especially in the low band.

Base-Station Antenna Mounting

Often a number of antennas have to be mounted on one tower. The top is the preferred position to obtain the height advantage and to avoid the tower structure distorting the antenna pattern. The lower antennas are side-mounted on the tower, with the ones needing better coverage placed higher.

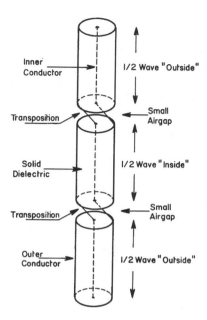

Inner Conductor

1/2 Wave "Outside"

Transposition

Small Airgap

Solid Dielectric

1/2 Wave "Inside"

Transposition

Small Airgap

Outer Conductor

1/2 Wave "Outside"

Fig. 2-12 Coaxial Collinear Array

Distance of antenna from tower structure. When an antenna is side-mounted on a tower, the tower's structure may affect both the horizontal pattern and the antenna impedance. To reduce these effects, the antenna distance from the tower should be a minimum of one wavelength. The larger the face of the tower structure in wavelengths, the farther the antenna should be mounted.

Distance between antennas on a tower. When a number of antennas are side-mounted on a tower, vertical separation is used to minimize the coupling between them. Recommended minimum vertical separation at 160 MHz is 20 ft, 10 ft at 460 MHz and 6 ft at 850 MHz. The vertical separations are measured from the tip of one antenna to the base of the one above it.

Antenna Lightning Protection

For lightning protection use antennas that provide direct dc grounding rather than those with a gap. The transmission line of the top-mounted antenna should be grounded to the top and bottom of the tower. The transmission line should also be grounded at the point where it enters the radio equipment housing. The antenna tower should be grounded properly to provide a low impedance for the lightning to ground. This is described in more detail in Chapter 14.

Wind Velocity Rating of Base-Station Antennas

This rating is defined as the maximum wind velocity that an antenna assembly can withstand without physical damage. Acceptable ratings vary according to location in three wind loading zones in the continental United States. The ratings in table 2-1 and the map of the wind loading zones in figure 2-13 are from NIJ-STD-0204.00.

TABLE 2-1 WIND VELOCITY RATING

| Antenna Base Height Above Ground | Wind Loading Zone | | |
| | A | B | C |
		Wind Velocity km/hr/(mi/hr)	
Less than 90 m (295 ft)	114 (71)	132 (82)	145 (90)
90-200 m (295-656 ft)	123 (76.5)	144 (89.5)	161 (100)
More than 200 m (656 ft)	145 (90)	168 (104)	193 (120)

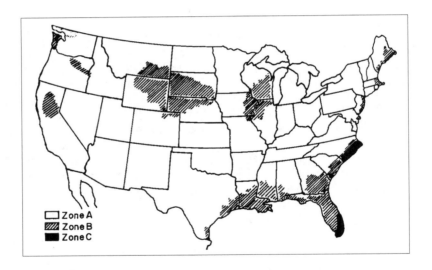

Fig. 2-13 Location of Wind Loading Zones

2.7 ANTENNA TOWERS

Antenna towers form an important part of most land mobile radio systems.

Types of Antenna Towers

There are two basic types of antenna towers: the self-supporting tower and the guyed tower, where the tower is supported by guys, or wires under tension.

Structural Standards

EIA Standard RS-222-C, "Structural Standards for Steel Antenna Towers and Antenna Supporting Structures," covers both self-supporting and guyed towers. Refer to this standard for details.

Loading

The design load of an antenna tower includes the combination of the tower and its appurtenances' dead weight, wind loading, and ice loading.

Wind loading. EIA Standard RS-222-C details the minimum wind loading. For example, a 250-ft tower is located in New York City. Figure 2-13 shows that New York City is in zone A. Referring to EIA RS-222-C, the antenna structure when fully loaded should be designed for a minimum horizontal wind pressure of 30 lb/ft^2 on flat surfaces.

Ice loading. Solid-ice loading is 56 lb/ft^2.

Grounding the Tower

There are two types of ground in an antenna tower. The primary ground is a conducting connection between the tower and earth or to some conducting body that serves in place of the earth. All towers must be grounded directly to a primary ground to be protected from lightning, high voltage, or static discharges. The secondary ground is a conducting connection between a tower appurtenance and the tower. All equipment on a tower must be grounded to the tower.

Minimum grounding. Minimum grounding consists of two 5/8-in.-diameter ground rods or equivalent driven at least 8 ft into the ground. The rod is bonded to a lead not smaller than No. 6, copper-connected to the nearest leg or to the metal base of the tower.

Grounding for self-supporting towers more than 5 ft in base width. One ground rod per leg is installed as described for minimum grounding.

Grounding for guyed towers. One ground rod is installed at each guy anchor and connected to the guy anchor by a lead not smaller than No. 6 copper.

Climbing facilities. Items such as fixed ladder and step bolts are designed to permit tower access.

Climbing safety devices. Climbing safety devices, such as safety belts, minimize falls.

Working facilities. Work platforms and access runways are working facilities.

Tower Lighting Requirements

Part 17 of the FCC Rules and Regulations, "Construction Marking and Lighting of Antenna Towers," gives details of tower lighting. These are also specified in the same way in "Obstruction and Lighting" from the U.S. Department of Transportation, Federal Aviation Administration.

2.8 RADIO-FREQUENCY TRANSMISSION LINES

In land mobile radio systems, most RF transmission lines are 50-Ω coaxial cables. The RF current is carried on the outside of the inner conductor and the inside of the outer conductor. Because of this, the outer conductor can be grounded on its outer surface. The insulating or dielectric material can be one of three types: solid dielectric, air dielectric, or foam dielectric. Cables are often classified according to these types of dielectrics.

Solid Dielectric Cables

In solid dielectric cables the insulating material between conductors is a solid material, usually solid extruded polyethylene dielectric. The inside conductor consists of stranded copper wire, while the outside concentric conductor is made of braided copper. These cables are low in cost but relatively high in losses. The losses are too excessive for runs on an antenna tower. They are used extensively to connect various RF test equipment, such as signal generators.

If a solid dielectric RG type is to be used, specify the AU suffix since this type resists moisture, sunlight, and corrosion. For example, the RG-8AU is better type RG-8.

Air Dielectric Cables

Air is the dielectric material between conductors in air dielectric cables. Insulators at intervals maintain the proper spacing between the inner and outer conductors. Both conductors are usually copper tubing. The space between the conductors is kept free of moisture by pressurizing it with dry air or nitrogen, which can be a burdensome maintenance problem. On the other hand, air dielectric cables have relatively low losses. When long runs or high frequencies are involved, you have to weigh the air dielectric's maintenance problems and its lower losses against the foam dielectric. Newer types of foam dielectric cables have losses very close to those of air dielectric cables.

Foam Dielectric Cables

In foam dielectric cables the dielectric material between conductors is a foam plastic dielectric, usually polyethylene foam. Pressurization is not required to keep out moisture because the space between conductors is completely filled. The inner conductor is usually a copper tube, while the outer conductor is usually a thin-walled corrugated copper tube.

RF Cable Attenuation

Table 2-2 gives an idea of typical attenuation in dB per 100 ft of different RF coaxial cables in the various mobile radio system bands. For exact attenuation of specific cables, consult manufacturers' catalogs.

TABLE 2-2 TYPICAL APPROXIMATE ANTENNA CABLE ATTENUATION (DB PER 100 FT)

Type of Cable	Frequency (MHz)			
	30	160	450	850
Solid dielectric, RG-8	1.0	2.9	5.2	—
1/2-in. foam	0.45	0.9	1.6	2.4
7/8-in. air	0.2	0.5	0.8	1.25
7/8-in. foam	0.23	0.5	0.9	1.4
1 5/8-in. air	0.15	0.25	0.45	0.64
1 5/8-in. foam	0.12	0.3	0.55	0.85

Preventing Transmission-Line Problems

Buried cables always should be jacketed to avoid corrosion problems in the soil if the outside metal of the cable is not protected by a plastic outer sheath. Moisture in a cable causes deterioration. To minimize this, waterproof the connector at the antenna input by taping the entire connection with low-temperature tape and then coating it with a sealant. Another method for waterproofing cable connectors is to use thick heat-shrinking tubing over the connector. Heat is applied to the tubing, and after cooling, the tubing is covered with a coating of sealant.

Detecting Transmission-line and Antenna Problems

A through-line RF wattmeter is very useful in checking the transmission line and antenna. This meter reads the transmitter output power from the transmitter to the antenna. It also can measure the reflected power from the antenna to the transmitter.

If the reflected power is 10% or more of the output power, a problem exists in the transmission line or the antenna. The transmission line then can be isolated and tested separately. A dummy load is used in place of the antenna and a portable transmitter instead of the base transmitter. A portable through-line RF power meter is used to measure the transmitter output and the reflected power. A reflected power of 10% or more of the output power then indicates a problem in the transmission line.

If the transmission line or associated connectors are not at fault, the antenna is probably defective. If the antenna is of the dc grounding type, it can be checked for continuity with the tower. If the antenna uses a star gap for lightning protection, it can be inspected for burn marks.

It is strongly recommended you include an installed through-line RF wattmeter capable of reading both output and reflected power when you purchase a base station. This enables the radio technician to make periodic measurements of the output and reflected power easily. A rising reflected power over a period of time indicates a growing problem in the transmission line or antenna.

Connectors

Some of the common connectors used in land mobile radio systems are

UHF. The UHF connector is simple to connect and is mechanically reliable, but the impedance is not constant. It should not be used above 300 MHz.

N. The N connector is more fragile and more difficult to install than the UHF type. However, it works well above 300 MHz to about 10,000 MHz. It has a constant impedance and is used extensively in the 400- and 800-MHz bands.

BNC. BNC is a small quick-disconnect connector with a bayonet-type lock coupling. The BNC is used on small-diameter cables for interconnecting within equipment.

2.9 CONSOLES

Dispatcher consoles control the radio system, permit the dispatcher to select and turn on different transmitters, turn on various audio tones to perform various functions, and so on. There are many different kinds and sizes of consoles. Some of the different functions and associated equipment are described here.

Volume Units (VU) Meter

A VU meter reads the volume of speech. It is a standardized meter with special electric and dynamic characteristics specified in American National Standard, "Volume Measurements of Electrical Speech and Program Waves," C16.5. The meter's sensitivity is adjusted so that zero volume units are read when the meter is connected across a 600-Ω resistor in which a 1000-Hz sine wave dissipates a power of 1 mW. Then, for a sine wave only, the meter is calibrated in decibels above or below 1 mW (dBm). One volume unit equals 1 dB for a sine wave only. This equality of volume unit with decibel unit is not true for nonsinusoidal inputs such as speech. Without compression or limiting, the instantaneous peaks of a zero VU voice signal equal the peaks of a sine-wave signal of +10 dBm. However, there is no conversion from VU to dBm, or vice versa.

Tone Generation

A number of different tone formats are generated in the console and transmitted to perform a number of signaling and activating functions. Some of these and their functions:

Single tones.   A single tone is one tone for a fixed time. It can be used as an alert tone preceding an important announcement. Often a 1000-Hz tone is used.

Two-tone sequential.   Two-tone sequential consists of one audio tone followed by another. It is often used for contacting pagers. A pager encoder that generates these tones is often incorporated in a dispatching console. The two-tone sequential system also is used for selective calling applications, such as contacting specific groups of vehicles. The vehicle decoder can activate a horn, for example. It is used to access tone-controlled remote repeaters and base stations, too.

Five-tone sequential.   In a five-tone sequential five tones are sent out one after another, each tone representing a digit. This also is used for pagers and for contacting groups of mobiles or walkie-talkies. It is a much faster system than the two-tone sequential. However, the two-tone sequential can be used with a tone alert followed by voice, whereas the five-tone sequential is used primarily in tone-alert-only systems. Both the two-tone and five-tone systems are described in some detail in Chapter 11.

Digital pulse code.   In a digital pulse code each digit in a number is represented by audio tone pulses. The specific code alerts the selected mobile receiver to an incoming call. One format uses a pulsed 1500-Hz tone and another uses 2805 Hz.

Dual-tone, multifrequency (DTMF).   DTMF is the touch-tone system used in pushbutton telephones and for various functions in two-way radio systems. This system uses two simultaneous tones for each push of the button. One of the two simultaneous tones is from a high group of frequencies, the other from a low group. Table 2-3 shows the format of the DTMF system. There is also a 16-digit DTMF format used in special security systems.

TABLE 2-3 DTMF FORMAT

		Buttons		
	697	1*	2	3
Low Group	770	4	5	6
Frequencies (Hz)	852	7	8	9
	941	*	0	#
		1209	1336	1447
		High Group Frequencies (Hz)		

* Pushing button 1 generates two tones simultaneously: 697 Hz in the low group of frequencies and 1209 Hz in the high group of frequencies.

Phone-Patch (Interconnect)

A phone-patch is used to connect the telephone land line system to the base station so that a mobile unit can communicate with someone at a telephone. The phone-patch is accomplished by a voice coupler and a hybrid circuit. The telephone company supplies the voice coupler, a connecting device attached to a telephone set, with a switch to connect and disconnect the coupler. The hybrid circuit is used to connect a two-wire circuit (the telephone voice coupler) to a four-wire circuit (two wires for the base radio receiver plus two wires for the base transmitter). The hybrid circuit prevents the base radio receiver audio from reaching the base radio transmitter's audio circuit. A simplified block diagram, figure 2-14, shows the basic idea.

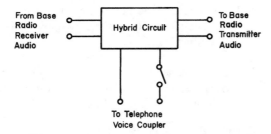

Fig. 2-14 Phone-Patch

Secode Modular Automatic Radio Telephone System (SMART) is a more sophisticated direct-dial phone interconnect that operates full or half duplex.

Lights and Controls

There are a number of lights and controls common to many consoles:

Monitoring voting receiver status. Voting receivers are used to increase coverage from mobile units to the base station. The mobile units transmit to strategically situated voting receivers in the service area. Dedicated telephone lines or microwaves send the audio from these receivers to a comparator at the dispatching center. The comparator picks out the receiver with the best signal and sends its "voted audio" to the console and/or repeater base station. At the console three lights for each voting receiver monitor its status. For example, a red light may mean out of service; a yellow light, voting; and a green light, voted receiver.

Busy light. A busy light indicates channel usage in consoles. When multiple consoles share a common radio channel, FCC regulations require that one user, the control point, be responsible for proper usage of the channel. The control point must have an indication of channel usage by others and a means of rendering the channel transmitter inoperative by other users.

Mute control. Different receiver channels can be muted: lowering volume to a low background level that can still be heard.

Call light. The call light flashes when receiver audio is coming in, even if the audio is muted.

Selected audio. In selected audio, audio is heard from a selected channel only.

Unselected audio. In unselected audio, audio is heard on a separate speaker from an unselected channel.

Transmitter controls and lights. A transmit on-off pushbutton with corresponding red and green lights is usually standard. However, foot switches are sometimes used instead of a pushbutton to turn on and off the transmitter.

Console Microphones and Headsets

Many dispatching consoles normally come with microphones. Headset connections in a dispatching console are sometimes special products (SP) or add-on kits.

Console microphones. Console microphones are often the dynamic or moving coil type of microphone, which requires amplification and is directional—you have to speak directly into the microphone. Console microphones are used primarily in dispatching centers with one console.

Headsets. Headsets have both earphones and a microphone that the dispatcher wears. In large dispatching centers headsets often are used because console loudspeakers from adjacent consoles would interfere with each other. Headsets are used also in separate telephone consoles in public safety dispatching centers such as police and fire. Headsets come in different types depending on the particular requirements at a dispatching center. Some of the different types and accessories:

One Ear or Both Ears.   The one-ear type allows the free ear to hear any verbal instructions.

On-the-Ear or In-the-Ear.   The on-the-ear headset uses a muff or cushion at the receiver portions. The element rests on the ear or contours the ear. The in-the-ear uses an "ear tip" plastic element inside the ear canal via an acoustic tube from the receiver transducer. This type of headset may irritate the sensitive ear canal. Many headset users prefer the on-the-ear type.

Amplifier/plug case.   The amplifier/plug case contains plugs for transmit-receive audio signals. An amplifier may be built in the plug case and fed by voltage supplied from the console to amplify the received signal. All microphones except the carbon type require an amplifier.

Quick disconnect.   A quick disconnect allows the headset to be disconnected from plug case and cord. The user need not remove the headset if he or she wants to move away from the console.

Clothing clip.   A clothing clip is used to relieve the weight of the cord from the headset.

Some Terms in Multiple Consoles

Multiple consoles are used in very large systems. Some terms to know:

Crosspatch.   Crosspatch is a control center subsystem that permits a mobile on one channel to communicate with one or more mobiles on a different channel through the control center console.

Patch intercom.   A patch intercom is a control center subsystem that allows a dispatcher at a remote console to communicate with a mobile unit on a different channel via the control console.

2.10 REMOTE CONTROL OF BASE STATION TRANSMITTERS

Dispatching centers may not be near an optimum antenna site. This may require remote control of the transmitter, turning it on and off and also changing frequencies. DC and tone are the two basic remote control methods.

DC Control

Dc control requires a dc connection from the control point to the remote base transmitter. Figure 2-15 shows a drawing of a dc control system.

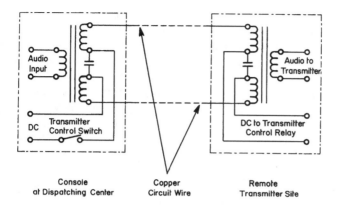

| Console at Dispatching Center | Copper Circuit Wire | Remote Transmitter Site |

Fig. 2-15 DC Remote Control of Transmitters

Tone Control Using Telephone Tie Lines

New copper lines that have a dc connection between two points are now difficult to obtain from a telephone company. Because of this, most land mobile radio systems use tones to turn on remote base transmitters and to change frequencies.

One popular method of controlling transmitters is to use two audio tones. Tone one, sent in a short burst at a high amplitude, sets up the transmitter for the second tone. The second tone, sent in a shorter time interval, selects the transmitter frequency. There is an audio tone for transmitter frequency F_1, another audio tone for transmitter frequency F_2, and so on.

The frequency selection tone is followed by tone one and voice. This time tone one is at a very low amplitude and stays on as long as the transmitter is on. The same tones that are used over telephone tie lines also can be used with RF control stations and microwaves to control remote base transmitters.

Frequency Jumping of Remote Base Transmitters

Mobile radio systems in many cities have experienced frequency-jumping problems. For example, a dispatcher at a large fire department was operating a fire on frequency F_3 at 1:00 a.m. Suddenly the tone-controlled remote transmitter jumped frequency, so without realizing it, he was operating on frequency F_1. It caused tremendous confusion in the middle of the operation: The fire vehicles lost communications with their dispatching center, and in an adjacent borough, communications were being blocked by the base transmitter, whose frequency had jumped. Such frequency jumping had happened to this fire department before—about once every two months on a random basis—but never during critical operating times. And, since the frequency jumping was infrequent and at random, it was very difficult to pin down.

Analysis and solution.   A four-frequency transmitter has four frequency control circuits, one for each frequency. The control circuits in this specific case were flip-flops, each with two states: on and off. The F_1 frequency selection tone turns on the flip-flop, which selects the F_1 frequency. Similarly, an F_2 selection tone turns on the flip-flop for the F_2 frequency, and so on. The equipment is designed so that if the power is turned off, the transmitter frequency selection reverts to F_1. This means that if there is a momentary power failure, even for milliseconds, the frequency jumps back to F_1.

The solution is to supply the flip-flop control circuits with a constant dc voltage. A small battery placed across the frequency-control-circuit dc supply prevents this type of frequency jumping. A diode in series with the battery prevents the battery from being drained while the regular dc voltage is on.

Steps to prevent frequency jumping:

1. Determine from the equipment instruction book if turning off the power causes the transmitter to revert to F_1. If this is so, proceed to step 2.
2. Determine where the frequency control circuits are. Locate the dc supply for these circuits.
3. Place an appropriate small battery, with a diode in series, across the dc supply for the frequency control circuit.

REFERENCES

U.S. Government Documents Without Listed Authors

Batteries for Personal/Portable Transceivers. NCJ 25994. NIJ-STD-0211.00, June 1975. GPO Stock No. 027-000-00342-7.[*]

Batteries Used with Law Enforcement Communications Equipment: Chargers and Charging Techniques. NCJ 10692. LESP-RPT-0202.00, June 1973. GPO Stock No. 027-000-00216-1.

Body-Worn FM Transmitters. NCJ 47378. NIJ-STD-0214.00, December 1978. GPO Stock No. 027-000-00711-2.

Continuous Signal-Controlled Selective Signaling. NCJ 71097. NIJ-STD-0219.00, August 1980. GPO Stock No. 027-000-01041-5.

Control Heads and Cable Assemblies for Mobile FM Transceivers. NCJ 77184. NIJ-STD-0216.60, 1982.

Fixed and Base Antennas. NCJ 77186. NIJ-STD-0244.01, 1981.

Fixed and Base Station Antennas. NCJ 41996. NIJ-STD-0204.00, November 1977. GPO Stock No. 027-000-00567-5.

* GPO refers to the stock number. These may be purchased from Superintendent of Documents, U.S. Government Printing Office, Washington, DC 20402.

Fixed and Base Station FM Receivers. NCJ 29643. NIJ-STD-0206.00, September 1974. GPO Stock No. 027-000-00283-8.

FM Repeater Systems. NCJ 41975. NIJ-STD-0213.00, November 1977. GPO Stock No. 027-000-00568-3.

Measured Vehicular Antenna Performance.[*] NIJ Report 201-85.

Mobile Antennas. NCJ 13319. NIJ-STD-0205-00, May 1974. GPO Stock No. 027-000-00250-1.

Mobile FM Receivers. NCJ 25996. NIJ-STD-0207.00, June 1975. GPO Stock No. 027-000-003443.

Mobile FM Transmitters. NCJ 15244. NIJ-STD-0202.00, October 1974. GPO Stock No. 027-000-00287-1.

Personal FM Transceivers. NCJ 47380. NIJ-STD-0209.00, December 1978. GPO Stock No. 027-000-0078-0.

RF Coaxial Cable Assemblies for Mobile Transceivers. NCJ 28496. NIJ-STD-0212.00, September 1975. GPO Stock No. 027-000-00357-5.

[*] Single copies may be obtained from National Bureau of Standards, Building 221, Room B157, Gaithersburg, MD 20899.

References

3

Avoiding Problems with Squelch Systems

Mobile land radios originally used squelch systems to prevent the noise with high-gain receivers from being audible. The first and most basic squelch system is the carrier-activated squelch.

Later, a continuous-tone-coded squelch system (CTCSS) was used. In this system a subaudible tone is transmitted with the voice message. The tone has to be decoded in the receiver before any audio is heard. This system has the advantage of keeping out any extraneous transmitters on the same frequency. To increase the number of squelch codes, another system using digital codes was developed.

These systems are described in some detail together with associated problems and solutions.

3.1 GOING NAKED WITH CARRIER SQUELCH

Figure 3-1 illustrates the basic method of carrier squelch. A high-pass filter at the detector output passes frequencies above the voice band. The filter output is essentially noise, which is then amplified and rectified into dc voltage. The dc voltage is applied to an audio stage, cutting off the audio signal. A variable resistor controls the dc voltage and thus the squelch level. In some cases a fixed resistor is used to set the level of squelch.

When a carrier signal is received, the limiter stages decreases the receiver gain. This lowers the noise, which in turn reduces the squelch voltage so that the receiver audio is activated.

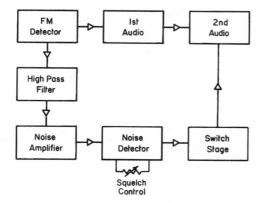

Fig. 3-1 Block Diagram of Carrier-Activated Squelch

Some Squelch Terminology

Squelch tail. Squelch tail refers to a burst of noise at the end of a transmission. The squelch tail results because the increase in receiver noise, caused by loss of carrier, must charge resistor-capacitor networks before squelch occurs. It takes a finite time for these networks to become charged. Special circuitry can be used to eliminate squelch tail, which is discussed later in this chapter

Tight squelch sensitivity. Tight squelch sensitivity is the minimum RF signal that unsquelches the receiver with the squelch control in the maximum squelched position.

Threshold squelch position. Threshold squelch position is the squelch control's position, starting from the maximum unsquelched position, that first reduces the audio noise power by 6 dB. This is sometimes referred to as 6 dBQ in receiver specifications.

Threshold squelch sensitivity. Threshold squelch sensitivity is the minimum RF signal that unsquelches the receiver with the squelch control in the threshold squelch position. Typical values are 0.2 μV.

Carrier squelch vulnerability. With carrier squelch the dispatcher hears the audio from all RF signals on the same frequency. This is particularly serious when communicating with emergency vehicles, and there can be confusion about who is dispatching. For example, a fire engine in one city dashed to an address after hearing a fire announced on the radio. However, the fire was at the same street address in another city. In addition, repeaters must not be used with carrier squelch. Extraneous RF signals turn on the repeater's transmitter if the repeater's receiver uses carrier squelch. These problems prompted the development of special coded types of squelch.

3.2 CONTINUOUS-TONE-CODED SQUELCH SYSTEMS

In CTCSS an encoder adds a specific continuous tone, from 67.0 to 250.3 Hz, to the transmitter modulation. The tones are below the audio bandpass of the radio, 300 to 3000 Hz, and thus cannot be heard.

How CTCSS Works

The associated receivers are equipped with a tone decoder that responds only to that specific tone and enables the receiver audio. Without this specific tone there is no receiver audio output, so the operators at the system's receivers do not hear nuisance communications on the same radio frequency. However, it should be noted that this does not eliminate co-channel interference—since while the receiver audio is enabled by the proper tone, co-channel interference can be heard.

Figure 3-2 shows a simplified diagram of CTCSS. The transmitter sends out a subaudible tone. In the receiver a low-pass filter passes the subaudible tone to a narrowband circuit that responds to only one specific subaudible tone. The detector changes the tone to a dc voltage, turning on the switch stage, which enables the audio. The high-pass filter in the decoder prevents any of the tones or their harmonics from entering the receiver's audio stages. The narrowband circuit can be a resonant reed or an active filter circuit.

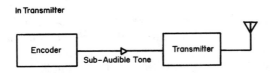

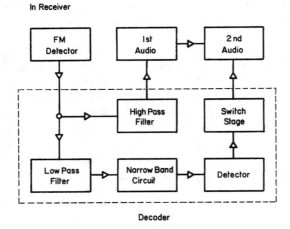

Fig. 3-2 Continuous-Tone-Coded Squelch System

It should be noted that a receiver that does not have CTCSS can still pick up a transmission that has CTCSS. However, since this receiver does not have the high-pass filter that is part of the CTCSS decoder, it picks up a hum in the background.

CTCSS is used so operating personnel do not have to listen to other radio systems on the same channel. It also is used within the same system where two or more radio transmitters share the same radio frequency, and when a number of different mobile users share a repeater.

CTCSS carries different trademark names by the various manufacturers of two-way radio systems. General Electric calls its Channel Guard, RCA has Quiet Channel, and Motorola has Private Line. In England it is sometimes referred to as "tone lock." In addition, tone equipment manufacturers sell CTCSS devices that can be added to two-way radio systems.

TABLE 3-1 ELECTRONIC INDUSTRIES ASSOCIATION CTCSS TONES (HZ)

Group A	Group B	Group C
67.0	71.9	74.4
77.0	82.5	79.4
88.5	94.8	85.4
100.0	103.5	91.5
107.2	110.9	
114.8	118.8	
123.0	127.3	
131.8	136.5	
141.3	146.2	
151.4	157.7	
162.2	167.9	
173.8	179.9	
186.2	192.8	
203.5	210.7	
218.1	225.7	
233.6	241.8	
250.3		

*By permission of the EIA.

CTCSS tones. The Electronic Industries Association specifies the tone frequencies in EIA Standard 220-B. The tones are divided into three groups: A, B, and C, as shown in table 3-1. It is advisable to use tones within one group for all radio frequencies within a system. This also applies to a shared repeater. The subaudible frequency for both transmitter and receiver can be selected by vibrating resonant reeds that transmit and respond to a specific tone. There are also tunable units that can be set at any of the CTCSS tones.

Disabling CTCSS. Part 90, Private Land Mobile Radio Services, FCC Regulations, Section 90.403c, states: "Licensees shall take reasonable precautions to avoid causing harmful interference. This includes monitoring the transmitting frequency for communications in progress and some other measures as may be necessary to minimize the potential for causing interference."

In shared-frequency situations the receiver CTCSS may have to be disabled temporarily to comply with this FCC regulation. When the CTCSS is turned off, the carrier-activated squelch system takes over.

Clipping problem. When CTCSS is installed, there is a delay of about one-third of a second between pushing the PTT button and speech reception. If operating personnel do not wait before speaking, the first word is clipped. This problem can be solved by training personnel to push the PTT and then pause momentarily before talking.

CTCSS carrier frequency deviation levels and problems. EIA Standard 220-B gives a range of CTCSS carrier frequency deviation from 500 to 1000 Hz for transmitters with a carrier deviation of 5 kHz. Some vendors' transmitters have an adjustable potentiometer for the CTCSS deviation, but others have a fixed resistor. If the CTCSS deviation is set too low, the received transmission is not heard. If it is set too high, a bleed-through hum occurs in the receiver when the subaudible CTCSS tone is strong enough to go through the high-pass filter in the receiver decoder. A CTCSS deviation of 750 Hz, or halfway between the EIA limits, is recommended. Sometimes, where there is a fixed resistor in the transmitter, the radio technician must be given tolerance limits. Experience has shown that a lower limit of 650 and an upper limit of 850 Hz works quite well.

Example of a problem with CTCSS deviation levels. In a large fire department, a dispatcher was transmitting a critical message to firefighters at a big fire. Several syllables dropped out of the middle of the message, causing considerable confusion. Attempts to repeat this dropout were unsuccessful even though the department used the same fire vehicle at the same location. A few weeks later a message from a fire vehicle suffered the same dropout in the middle of a sentence. This problem repeated itself on a random, intermittent basis but only in one borough. There was never a dropout in the other four boroughs.

An investigation revealed the problem borough was using a new type of voting receiver. The new receivers were slightly less sensitive than the older ones used in the four problem-free boroughs. The CTCSS carrier deviation in all the base and mobile transmitters had been set at 500 Hz or less. A slight change in multipath propagation then caused random intermittent dropouts at the receiver.[*]

The radio frequency deviation for CTCSS was adjusted between 650 and 850 Hz for all transmitters, which solved the problem.

CTCSS deviations levels for transmitters with 2.5 kHz carrier deviation. In the future the FCC may require that transmitters' carrier deviation be reduced from 5 kHz to around 3 kHz. EIA Standard 220-B also gives CTCSS carrier frequency deviation levels from 0.35 kHz to 0.6 kHz for transmitters with 2.5 kHz carrier deviation.

Eliminating squelch tail in CTCSS. In some systems a special circuit in the transmitter eliminates the squelch tail. At the end of each transmission this circuit advances the phase of the CTCSS's transmitted tone, which is emitted for a short time after the transmission. This is known as a reverse burst, and it cancels out the squelch tail. Some receivers now are designed in such a manner that the noise burst at the end of the signal is as short as 10 milliseconds (ms), which effectively eliminates the squelch tail, also.

3.3 KEEPING A CTCSS TONE EXCLUSIVELY YOURS

There are 37 different CTCSS tones. It is possible for two radio stations to have the same tone, thus nullifying its use. This is especially important in the public safety radio services. For example, a newly licensed volunteer fire radio station in Pennsylvania came on the air with a CTCSS tone of 131.8 Hz. One hundred and twenty-five miles away a fire radio station in Connecticut was operating on the same frequency with the same CTCSS. The Connecticut station could no longer communicate effectively until a change was made. Large radio systems are reluctant to make changes after they go on the air.

Addressing the Problem

There are two situations when CTCSS tones are duplicated: when a newly licensed station starts transmitting with a CTCSS tone and when a licensed station installs CTCSS.

Newly licensed station with CTCSS. To prevent duplication notify frequency coordinators in your area that you are operating with a particular tone. Organizations such as APCO now ask licensees for their proposed CTCSS tone. The distance involved should extend out to 200 miles away for high-band VHF. For UHF the distance can be reduced to about 150 miles, and to about 80 miles for the 800-MHz band.

Existing station switches to CTCSS. Make a list of all stations on your frequency by listening for a long time with CTCSS off. Notify all of these stations once a year

---

[*] This slight change in multipath propagation could be caused by a moving vehicle between the stationary fire vehicle and the voting receiver.

that you are operating with a particular CTCSS tone. Ask them to notify you if they plan to install CTCSS.

3.4 GOING UP THE LADDER WITH DIGITAL SQUELCH

Digital-coded squelch systems have been devised to obtain more coded channels than CTCSS. Unlike CTCSS, these are not compatible from one manufacturer to another.

In Transmitter

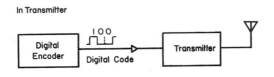

In Receiver

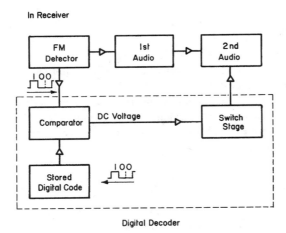

Digital Decoder

Fig. 3-3 Digital-Coded Squelch Systems

How Digital Squelch Works

In digital-coded squelch systems a specific binary code word is used instead of a subaudible tone. Figure 3-3 shows one method in simplified form. In this system the digital encoder generates a specific binary word consisting of bits. The bits are either logic "l" or logic "O." This binary word is then converted to a signal in which the logic l causes a positive shift of the transmitted frequency and the logic O causes a negative shift of the frequency.* When received, these frequency shifts are converted back to the original digital code. The re-

---

* This type of digital modulation requires a direct frequency-modulated transmitter instead of a phase-modulated transmitter.

ceived digital code is compared with a stored digital code in the digital decoder. If they are the same, a dc voltage output is produced to enable the audio.

Thus the digital-coded squelch works in a manner similar to CTCSS. Both are operated continuously, and both usually are operated so that when they are switched out of the circuit for monitoring or other purposes, a carrier-activated squelch circuit takes over.

The digital squelch encoder for the transmitter can transmit a reverse code similar in function to the reverse burst in CTCSS to eliminate the squelch tail.

The number of digital squelch combinations depends on the number of bits. For 8 bits the combinations are 2^8 or 256.

REFERENCES

Chaney, W. G., and R. T. Meyers. "New Designs for High Performance," *IEEE Transactions on Vehicular Communications,* March 1965.

Daniel, Jack. "Sub-audible Tone Installation," *Communications,* June 1979.

Hoag, T. E. "Interfacing the CTCSS Decoder," *Communications,* June 1980.

Roy, W. R. "Conversion of a VHF Communications System from Standard to Tone Squelch," *IEEE Transactions on Vehicular Technology,* May 1969.

Publication without listed author

Standard RS-220B, Electronic Industries Association, Washington, DC.

4

Combining Receivers and Transmitters into One Antenna

Antenna combiners are used in land mobile radio communications to allow one antenna to perform the functions of two or more. Antenna combiners for a number of receivers and transmitters are used more frequently with the advent of trunking, cellular, and other systems. Couplers using hybrid splitters, cavities, and isolators are described.

The duplexer is the most widely used antenna combiner. It couples a single transmitter and a single receiver into one antenna. This chapter describes the different types of cavity filters that are combined to form duplexers and discusses factors in selecting and using duplexers.

Both transmitter and receiver intermodulation in an antenna combiner is described together with prevention methods.

Receiver multicouplers are used at multifrequency receiving sites. This chapter discusses receiver multicoupler operation and some specification factors.

4.1 DUPLEXERS

A duplexer is a device that permits a transmitter and a receiver to use a single antenna simultaneously. The duplexer operates with two filters. One filter, between the receiver and the antenna, prevents the transmitter signal from desensitizing the receiver. A second filter, between the transmitter and the antenna, attenuates the transmitter noise.

Terms and Equipment to Know

Desensitizing.   The front end of a receiver by itself is not very selective. It responds to large off-receiver channel signals from the transmitter, but large signal voltages upset the receiver bias voltages, which decreases the receiver's sensitivity.

Transmitter noise.   Some of the transmitter power spreads out at a low level over frequencies well above and below the assigned transmitter frequency. This spread-out low-level transmitter radiation is called transmitter noise. Transmitter noise masks the receiver's desired signal, degrading receiver performance. It should be noted that transmitter noise cannot be filtered out at the receiver since it is on the receiver channel.

Cavity filters.   Cavity filters are used in duplexers and other antenna combiners. There are two basic types of cavities: the bandpass and the band-reject.

Bandpass Cavity.   The diagrams in figure 4-1 show the cavity itself, the functional equivalent, and the attenuation with frequency of the bandpass cavity. The cavity is one-

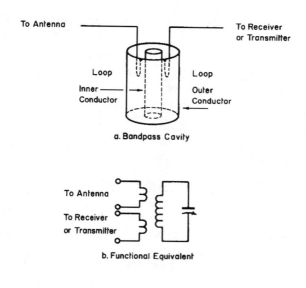

a. Bandpass Cavity

b. Functional Equivalent

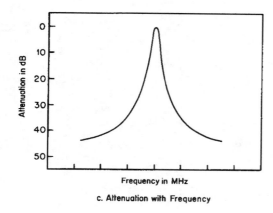

c. Attenuation with Frequency

Fig. 4-1 Bandpass Cavity

quarter of a wavelength long. The inner conductor can be raised or lowered to tune the center frequency of the cavity. The position and size of the two loops determine both the output frequency response and the insertion loss. The narrower the output frequency response, the greater the insertion loss.

Band-reject cavity.   The diagrams in figure 4-2 illustrate the cavity, the functional equivalent, and the attenuation with frequency of the band-reject cavity. The cavity is similar to the bandpass cavity except there is only one loop instead of two and a narrow band of frequencies is rejected. This type of cavity also is known as a notch filter. Coupling the loop to the cavity determines the depth of the notch.

Types of duplexers.   There are three general types of duplexers: bandpass, band-reject, and combinations of the two.

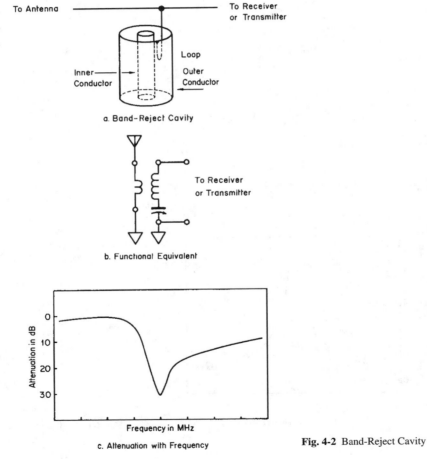

a. Band-Reject Cavity

b. Functional Equivalent

c. Attenuation with Frequency

Fig. 4-2 Band-Reject Cavity

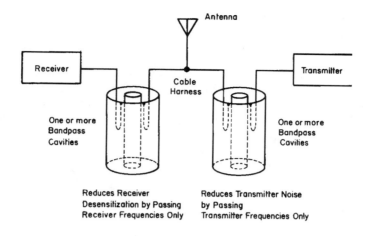

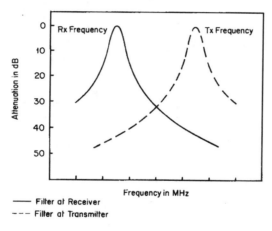

a. Bandpass Duplexer Circuit

b. Bandpass Duplexer Frequency Response

Fig. 4-3 Bandpass Duplexer

Bandpass Duplexer. Figure 4-3 shows the two basic cavities in a bandpass duplexer. One bandpass cavity is placed between the receiver and the antenna to filter out the transmitter frequency and reduce receiver desensitizing. This filter has a bandpass that allows the receiver frequencies to pass. The other bandpass filter is placed between the transmitter and the antenna. This filter has a bandpass that reduces transmitter noise by passing only the transmitter frequency. The cable harness interconnects the cavities in the two sections of the duplexer. In addition, by using special lengths of cable in the harness, it acts as a matching device so the outgoing energy from the transmitter goes to the antenna

Sec. 4.1 Duplexers

as the path of least resistance. Also, the incoming signal from the antenna sees the receiver as the path of least resistance. The duplexer manufacturer supplies the cable harness, and the lengths should not be changed.

Band-Reject Duplexer.   Figure 4-4 shows the two basic cavities in a band-reject duplexer. One band-reject cavity is placed between the receiver and the antenna to reduce receiver desensitization by rejecting the transmitter frequency. The other band-reject cavity, placed between the antenna and transmitter, rejects the transmitter noise that is on the receiver frequency.

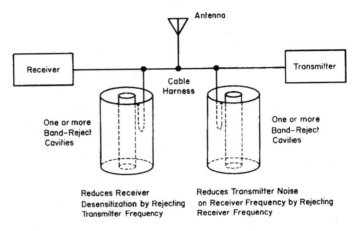

a. Band-Reject Duplexer Circuit

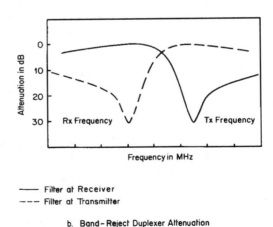

b. Band-Reject Duplexer Attenuation

Fig. 4-4 Band-Reject Duplexer

Combination Duplexer. Special coupling techniques make a single cavity act as a bandpass filter at one frequency and as a band-reject filter at another. This cavity, placed between the receiver and antenna, is used to pass the receiver frequency and reject the transmitter frequency. A second cavity, between the transmitter and antenna, is used to pass the transmitter frequency and reject the transmitter noise at the receiver frequency.

Using a duplexer properly.
The duplexer must not be used above its rated power. The duplexer must be designed to operate at, or less than, the frequency separation between receiver and transmitter. The duplexer is usually rated at a certain minimum frequency, such as "5 MHz or more."

The duplexer must provide sufficient isolation to prevent receiver desensitization and reduce transmitter noise. Manufacturers supply attenuation curves that can be compared.

The insertion loss must be small enough not to affect system performance significantly. A 3-dB insertion loss represents a power loss of one-half. Again, manufacturers supply data sheets that can be compared.

Do not change lengths of cables, types of cables, or connectors in duplexer cable harnesses. Radio technicians have made these changes without realizing they were changing the duplexer's performance characteristics.

4.2 REDUCING INTERMODULATION IN ANTENNA COMBINERS

Intermodulation is mixing two or more radio frequencies in a nonlinear device to produce another frequency. It is an important consideration in antenna combiners.

Understanding Intermodulation

Examples of two frequency intermodulation products are

$$2A \pm B = C; \quad \text{third order}$$
$$3A \pm B = C; \quad \text{fourth order}$$
$$4A \pm B = C; \quad \text{fifth order}$$

A is the frequency of one source and B is the frequency of another source. Coefficients represent harmonics; 2 is the second harmonic, 3 is the third harmonic, and so on. The sum of the coefficients is the order of intermodulation and C is the intermodulation product. Only the odd-order difference products are significant in land mobile radio systems—that is the third, fifth, and so on, order difference products. The most important intermodulation product is the third order, represented by the equation $2A - B = C$.

Transmitter and receiver intermodulation products are produced differently and require different solutions.

Transmitter intermodulation.
Figure 4-5 illustrates transmitter third-order intermodulation in an antenna combiner. The second harmonic of one transmitter combines

with the fundamental frequency of another to produce a difference frequency. This third-order intermodulation is radiated out to cause interference.

There are higher orders of intermodulation products than the one illustrated in figure 4-5. However, a ferrite isolator can be used to eliminate the effects of all transmitter intermodulation products.

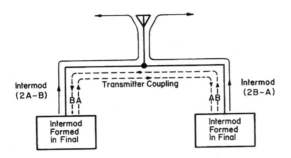

Fig. 4-5 Transmitter Intermod in Antenna Combiners

Ferrite Isolators. Figure 4-6 illustrates ferrite isolator operation. Radio frequency in the isolator can go in only one direction, which prevents undesired signals from coming down the antenna. The isolator itself generates a second harmonic. A filter is required at port 2 in figure 4-6 to reduce this second harmonic. In antenna combiners the filter may be a bandpass cavity. The isolator has an insertion loss of 0.5 dB per unit. Each isolator reduces the undesired signal by 25 dB. It should be noted that isolators decrease signals coming from the antenna, so you cannot combine receivers with transmitters using isolators because the isolator reduces the signal to the receiver.

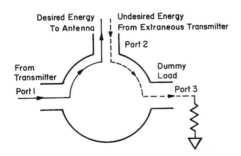

Fig. 4-6 Ferrite Isolator

Receiver intermodulation. In receiver intermodulation the RF signals from two transmitters mix in the front end of a receiver to produce an intermodulation product. In antenna combiners the transmitters are on the same tower as the receivers, which makes receiver intermodulation an important factor. To reduce receiver intermodulation, use a bandpass cavity between a receiver and the antenna. A transmitter whose frequency is outside the bandpass frequencies has its signal attenuated. This helps prevent receiver intermodulation from becoming a problem.

4.3 COMBINING SEVERAL RECEIVERS AND TRANSMITTERS INTO ONE ANTENNA

Antenna combiners vary from diplexers to systems involving 20 or more transmitters and receivers. Diplexers are combiners for either two transmitters or two receivers.

Why Combine?

Often a number of different radio systems share a common radio site located in a favorable place, such as a mountaintop. An antenna combiner eliminates a multitude of antennas by sharing a common antenna on top of the tower. This gives all the systems better performance.

There may be as many as 20 transmitters and 20 receivers in one trunking system. The number of antennas required without an antenna combiner would not be practical on one tower. To solve this problem, 20 transmitters are coupled together, as are 20 receivers. The final stage of the combiner couples the combined transmitters with the combined receivers into one antenna.

General types of combiners.   Cavities, isolators, and hybrid splitters are used to combine a number of transmitters and receivers.

Cavity Combiners.   The bandpass/band-reject combiner is the most popular cavity type. It is used to couple a number of transmitters and receivers into a single antenna. Figure 4-7 illustrates in symbolic form a bandpass/band-reject combiner for four frequencies. The system can be expanded by adding a band-reject cavity for frequency F_4, a bandpass cavity for frequency F_5, and so on.

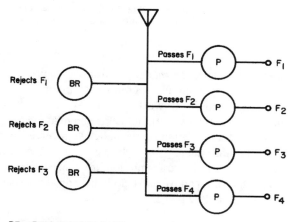

BR = Band-Reject Cavity Filter
P = Bandpass Cavity Filter
F = Frequency

Fig. 4-7 Bandpass/Band-Reject Combiner

Recommended minimum separations in megahertz between two adjacent channels are 0.5 for VHF low-band, 1.0 for VHF high-band, 2.0 for 450-MHz band, and 5.0 for 800-MHz band.

Cavity-Ferrite Combiner. A cavity-ferrite combiner is used to combine a number of transmitters into a single antenna. Ferrite isolators are used with bandpass cavities in figure 4-8 to control intermodulation.

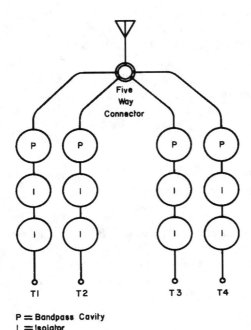

P = Bandpass Cavity
I = Isolator

Fig. 4-8 Cavity-Ferrite Combiner for Four Transmitters

Strip-Line Hybrid Coupler. The term *hybrid* derives from the hybrid audio circuit used in telephone work to connect a two-wire circuit to a four-wire circuit. Strip-line transmission lines are used to do the same thing in RF work. The four-wire circuit consists of two receivers or two transmitters, and the two-wire circuit is the antenna transmission line. The hybrid circuit supplies about 40 dB of isolation between the two receivers or transmitters, provided the hybrid load matches the antenna. The hybrid coupler shown in figure 4-9 has an insertion loss of about 4 dB. The strip-line transmission line consists of a conducting strip with a grounding conducting plane both above and below the strip. There is insulating material between the strip and both grounding planes.

Hybrid-Isolator Combiner. Figure 4-10 illustrates a four-transmitter combiner using hybrid splitters and isolators.

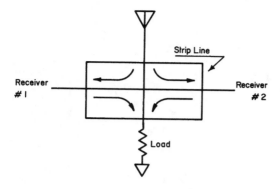

Fig. 4-9 Strip-Line Hybrid Coupler

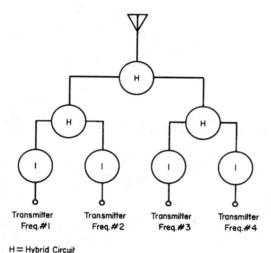

H = Hybrid Circuit
I = Isolator

Fig. 4-10 Hybrid-Isolator Combiner

4.4 RECEIVER MULTICOUPLERS

Receiver multicouplers include a bandpass filter between the antenna input and the antenna to prevent overloading by strong signals. A radio-frequency amplifier is used to compensate for divider losses. A broadband power divider divides the signal received from the antenna into equal parts while isolating the output ports from each other. Each output port feeds a receiver or a 50-Ω load if any output port is not being used. Figure 4-11 shows a diagram of a receiver multicoupler using hybrid splitters.

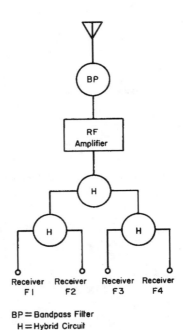

BP = Bandpass Filter
H = Hybrid Circuit

Fig. 4-11 Receiver Multicoupler

Receiver Multicoupler Performance Specifications

Noise figure. The noise figure is essentially the ratio in dB of the noise output divided by the noise input. A typical value for a receiver multicoupler in the VHF high-band and 450-MHz band is 6.5 dB.

Voltage standing-wave ratio (VSWR). The VSWR is a measure of the impedance match of the multicoupler's input and output to 50 Ω. A typical VSWR for a receiver multicoupler is 1.5 for both input and output.

Isolation. A signal at receiver frequency 1 will be attenuated by a certain number of dB at the port for receiver frequency 2. A typical isolation specification is 25 dB.

Preselector bandwidth. The preselector bandwidth is measured at the frequencies where the attenuation is 3 dB. A typical preselector bandwidth for VHF high-band is 1 MHz; for 450-MHz, 6 MHz.

Third-order intermodulation specification. In third-order intermodulation specification a typical performance specification for $(2A - B)$ is 80 dB below the input levels of frequencies A and B.

REFERENCES

Barrow, W. L., and W. W. Mieher. "Natural Oscillation of Electrical Cavity Resonators," *Proceedings of the IRE.* April 1940.

Howe, Harlan, Jr. *Strip-Line Circuit Design.* Dedham, MA: Artech House, Inc. 1974.

Klein, Robert J. "An Eight Transmitter Multiplexer," *IEEE Transactions on Vehicular Communications.* March 1965.

Roberts, Roy W., and James E. Lahey. "Ferrites for VHF Frequencies," *IEEE Transactions on Vehicular Communications.* March 1965.

Skomal, E. N. "Theory of Operation of a 3 Port Y Junction Ferrite Circulator," *IEEE Transactions on Microwave Theory and Techniques.* March 1963.

Uenishi, Kiyoshi, Kinichiro Arran and Hideo Ishi. "Transmitter Multiplexing System in UHF Mobile Radio," *IEEE Transactions on Vehicular Technology.* March 1965.

5

Controlling Frequencies

This chapter examines the FCC frequency tolerances of various frequency bands. Crystals are described, including their manufacture, the equivalent circuit, and series and parallel resonant circuits.

The frequency synthesizer is replacing many crystals in multifrequency radios. Frequency synthesizer components are examined, including the phase-locked loop and programmable devices.

The last subject covered in this chapter is a special frequency standard for a particular purpose.

5.1 KEEPING WITHIN FREQUENCY TOLERANCE LIMITS

FCC regulations require the carrier frequency of a transmitter be maintained within a percentage of the assigned frequency. This percentage varies with the frequency band, output power, and whether the transmitter is fixed or mobile.

Specific Tolerance Limits

Section 90.213 of Part 90 of the FCC regulations lists the tolerances for land mobile transmitters. They are summarized here by frequency band:

VHF low-band. The frequency tolerance for VHF low band is ±0.002% for all transmitters over 2 W and ±0.005% for mobiles of 2 W or less.

VHF high-band. The frequency tolerance for VHF high-band is ±0.0005% for all transmitters over 2W and ±0.005% for mobiles of 2W or less. A station operating on a 5-kHz channel is an exception: here the tolerance is ±0.0002% for all transmitters.

UHF (450 to 512 MHz) band.   The frequency tolerance for UHF is ±0.00025% for fixed or base stations and ±0.0005% for all mobiles.

806- to 930-MHz band.   The frequency tolerance for 806 to 930 MHz is ±0.00015% for fixed or base stations and ±0.00025% for mobile transmitters.

In land mobile radio systems, frequency control is basically one of two methods: quartz crystals for each frequency used or frequency synthesis of a number of frequencies based on one reference crystal.

Proposed FCC frequency tolerance limits.   The FCC has proposed that in the future the frequency tolerances be changed from percentages to parts per million. This would mean converting the present 806- to 930-MHz band percentage tolerances to ±1.5 parts per million for fixed or base stations and ±2.5 parts per million for mobile transmitters. The FCC has proposed that the frequency tolerances in parts per million be changed in the other bands since there will be channel division. Details will be given in the new Part 88 that will be published after hearings and comments. Part 88 is scheduled to come out in 1994 or early 1995.

5.2 CONTROLLING FREQUENCIES WITH CRYSTALS

Crystals for radio-frequency control are made of quartz. In many cases cultured quartz is grown in an autoclave, a large cylinder pressure vessel that can withstand extreme pressure and heat. The pressure range can be as high as 30,000 lb/in^2 and the temperature as high as 400°C. The nutrient is natural quartz material that grows on single-crystal seeds over a period of 30 to 45 days. The results are pure quartz bars from which crystals are cut.

Piezoelectric effect.   When crystal plates are cut in certain definite relationships to the axis of the crystal, it creates a piezoelectric effect: an electric field deforms the crystal plate, and a plate deformation results in an electric potential field on opposite sides of the crystal.

Equivalent circuit of a crystal.   The mechanical vibration system of the crystal has a mathematical equivalent in electrical systems. For example, an inductance can represent the crystal's mass inertia. Similarly, the crystal's elasticity is equivalent to a capacitor in series with that inductance. A resistor can represent a crystal's dissipation of heat. Figure 5-1 shows the equivalent circuit of the crystal. The output capacitor of this circuit is the crystal's electrostatic capacitance. The output capacitor also includes the capacitance of the crystal leads.

Crystal frequencies.   A quartz crystal plate has a number of mechanical resonances. The crystal vibrates at a frequency dependent on dimensions and crystal cut. At frequencies below 1 MHz the length-to-width ratio is important in frequency determination. At frequencies above 1 MHz the thickness determines frequencies. A thinner crystal oper-

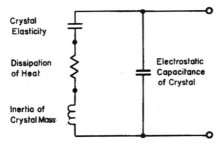

Fig. 5-1 Equivalent Circuit of a Crystal

ates at higher fundamental frequencies. For very high frequencies a crystal can be operated at one of its overtones. The overtone is a mechanical phenomenon and its frequency differs from the harmonic because of the crystal's mechanical loading. The harmonic is an electrical phenomenon and an exact multiple of the fundamental frequency. The crystal can be ground especially for overtone operation, and operation on the third overtone is used often. The angle of the crystal's cut, which is designated by letters, determines the range of frequencies. The AT cut is used often for frequencies from 550 kHz to 55 MHz.

Series and parallel resonance. A crystal can be made to operate in the parallel resonance mode by calibrating it with a small parallel capacitor. This circuit has a high output impedance, as figure 5-2 shows.

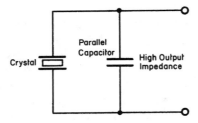

Fig. 5-2 Crystal Parallel Resonance

The crystal can be operated at series resonance by using a capacitance in series, as figure 5-3 illustrates. The series resonant circuit has a low output impedance.

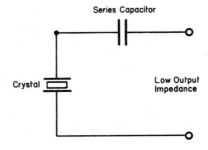

Figure 5-3 Crystal Series Resonance

Channel element.   Crystals in land mobile radios are packaged with other components to form a plug-in unit called a channel element. A typical channel element includes a crystal, an adjustable capacitor, thermistors, and a transistor. A thermistor is a device whose resistance decreases with a rise in temperature. This effect is used to compensate for the crystal's increase in frequency with temperature. The adjustable capacitor is set at a reference temperature of 25°C (77°F) for the desired frequency. The temperature-compensating network is designed to keep a transmitter within the FCC frequency tolerances over a specified temperature range of –30 to +60°C (–22 to +140°F).

5.3 FREQUENCY SYNTHESIZERS

Frequency synthesizers have become increasingly popular in land mobile radio systems, which employ a number of frequencies.

How Frequencies Work

The frequency synthesizer uses one reference frequency crystal and generates a number of different frequencies. Thus, instead of using one crystal for each transmitter frequency and one crystal for each receiver frequency, the frequency synthesizer produces all the required frequencies using only one reference crystal. The phase-locked-loop frequency generator is the heart of the frequency synthesizers.

Phase-locked-loop generator.   Figure 5-4 shows a simplified diagram of phase-locked loop (PLL). It consists of a VCO, a frequency divider, a phase detector, and a reference frequency.

The VCO is a self-excited RF oscillator whose frequency depends on the dc voltage the phase detector supplies to it. The VCO frequency is divided by the frequency divider.

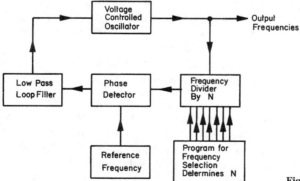

Fig. 5-4 Phase-Locked-Loop Generator

The frequency divider is a programmable digital frequency counter. Each time it counts a predetermined number N of RF cycles from the VCO, it delivers one output pulse to the phase detector, producing a train of pulses.

A reference frequency is generated by a crystal oscillator whose frequency is divided down, producing another train of pulses.

The phase detector compares the reference frequency oscillators and VCO's pulse trains. If there is a difference, the phase detector produces a voltage that varies the VCO frequency. When the two pulse trains' frequencies are equal and the phase difference is 90°, a dc voltage is produced, as shown in figure 5-5. The dc voltage holds the VCO on the desired frequency.

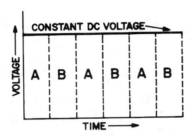

Fig. 5-5 Phase Lock of Two Pulse Trains, Same Frequency and 90° Out of Phase

Chirps from frequency synthesized transmitters. Chirps are transient transmissions at a rapidly changing frequency that may extend a few megahertz from the carrier frequency. These brief chirps can interfere with other uses, particularly television receivers operating in adjacent bands and other licensees operating digital systems. The FCC has proposed that all transmitters accepted under Part 88 (formerly Part 90) limit chirps to less than 20 milliseconds' duration.

Frequency selection. Figure 5-6 shows a diode matrix to illustrate frequency selection. This is a simple device that selects the proper N divider for a particular frequency in the radio. For example, when F_1 is selected the fused diode connected to the F_1 line is activated. In figure 5-6 this is the diode connected to the $N1$ line. All the other diodes connected to the F_1 line have had their fuses blown. A heavy current programs the diode matrix by blowing the fuses in the unwanted N divider lines.

The diode matrix itself has been replaced by programmable devices such as the programmable read-only memory, erasable programmable ROM, and electrically erasable programmable ROM.

Programmable Read-Only Memory (PROM). The PROM is used to perform the frequency selection. One form of PROM is essentially the diode matrix in a VLSI or chip. The fuse is a material such as polysilicon, which is part of the chip. To store the desired frequency information, heavy currents are used to blow the fuses and eliminate the frequencies that are not being used. If an additional frequency is added later, a fresh PROM must be used and "burned in" all over again. Some PROMs use transistors instead of diodes, but the principle is the same. Figure 5-7 illustrates the use of transistors in a PROM.

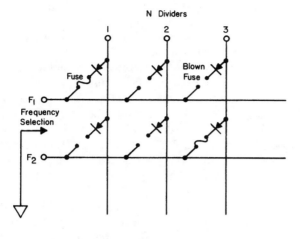

Fig. 5-6 Diode Matrix

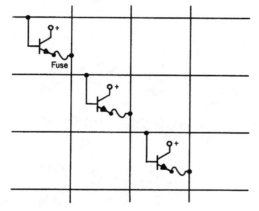

Fig. 5-7 PROM with Transistors

Erasable Programmable ROM (EPROM). The EPROM is a ROM in which ultraviolet light floods information in all memory cells simultaneously and erases it. New information is then electrically written into each cell. The information may be retained for 10 years or more, but it will gradually leak out. A floating-gate avalanche-injection metal-oxide semiconductor (FAMOS) is an example of one type of memory cell used in the EPROM.

Electrically Erasable Programmable ROM (E²PROM). The information in all cells is simultaneously erased electrically in E²PROM. Metal-nitride oxide semiconductor (MNOS) is an example of a cell used in E²PROM.

5.4 USING SPECIAL FREQUENCY STANDARDS FOR SPECIAL PURPOSES

Communications equipment used in some mobile radio systems requires special frequency control. One such system is simulcasting, where a number of base transmitters operate on the same frequency. A difference of 300 Hz or more between two transmitters can result in an audible heterodyne frequency in a mobile receiver. This occurs in areas where one base station signal is not strong enough to capture the other. The frequency tolerance of all base transmitters must be held to below 150 Hz to avoid the audible heterodyne frequency.

Ordinary crystal oscillators do not meet such strict tolerances. For example, at 450 MHz the FCC requires a frequency tolerance of ±0.00025% or ±1125 Hz. This can produce a heterodyne frequency of 2250 Hz between two base transmitters. The result could destroy communications in certain areas. A high-stability crystal oscillator can solve this problem.

High-stability crystal oscillators. Frequency tolerance of ±0.000005% is available commercially at 450 MHz, or ±22.5 Hz. Frequency tolerances this close call for special measures, including very precise grinding of the crystals and ovens that keep the crystals at a specific temperature.

Atomic frequency standards. The ultimate frequency standard uses resonance phenomena in atoms or molecules of cesium, rubidium, or hydrogen. For example, cesium has a natural frequency of 9,192,631,770 Hz. Atomic frequency standards are used in satellite navigation systems, which provide position information for vehicles in land mobile radio systems. The National Bureau of Standards (NBS) maintains a frequency standard based on the atomic natural resonance of cesium. The frequency standard is stable to 1 part in 10^{12}.

WWV frequency standard. The WWV frequency is a broadcast radio signal that can be used to calibrate frequency meters. The WWV signals are controlled by crystal oscillators that are offset from the NBS atomic frequency standard by a small known amount. WWV, located in Fort Collins, Colorado, broadcasts on 2.5, 5, 10, 15, 20, and 25 MHz. The NBS also has other frequency standards radio stations: WWVH in Maui, Hawaii, that broadcasts on 2.5, 5, 10, and 15 MHz, for one.

REFERENCES

Manassewitsch, Vadim. *Frequency Synthesizers, Theory and Design,* 3rd Ed. New York: John Wiley & Sons, Inc., 1987.

Ramsey, Norman F. "Past Present and Future of Atomic Time and Frequency," *Proceedings of the IEE 79.* No. 7. July 1991: 921–926.

———. "History of Atomic Clocks," *Journal of Research of National Bureau of Standards.* Vol. 88. 1983 301.

Salt, David. *HY-Q Handbook of Quartz Crystal Devices.* New York: Van Nostrand Reinhold Company, Inc., 1987.

6

Improving and Extending Area Coverage

Techniques to improve and extend area coverage in land mobile radio systems using one pair of frequencies have been developed. These are voting receiver, transmitter steering, simulcasting, and digital computer-controlled wide-area systems.

This chapter calculates the theoretical improvement in talk-back capability with a number of voting receivers. Two different types of voting systems are discussed together with some practical problems encountered in each. Transmitter steering using voting systems is explained.

The advantages and disadvantages of simulcasting are weighed, and a digital system that uses a computer to select receivers and transmitters to cover a large area is examined.

6.1 IMPROVING TALK-BACK CAPABILITY

The base transmitter in a land mobile radio system has much more power than mobiles or portables. In addition, the base antenna is much higher and usually more efficient than mobile or portable radio antennas. For these reasons the mobiles' and portables' communications limitation is their talk-back capability.

Using Receiver Voting Systems

One way to improve the talk-back capability is to use a receiver voting system, as shown in figure 6-1. A number of radio receivers located in strategic areas receive the RF signal from a mobile or portable unit. The detected audio then is sent to a comparator by telephone tie lines or microwave link. The comparator is usually located in a dispatching center. The "best" signal is voted in the comparator and the dispatcher hears that voted audio.

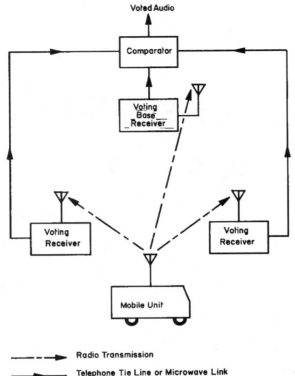

Voted Audio

Comparator

Voting Base Receiver

Voting Receiver

Voting Receiver

Mobile Unit

- - - ▶ Radio Transmission

——▶ Telephone Tie Line or Microwave Link

Fig. 6-1 Voting Receiver System

Three lights for each voting receiver indicate its status. A green light indicates the receiver has been voted the best. Another color light indicates the receiver is voting but has not been selected (this color varies depending on the system). A red light indicates the receiver is out of service. A reject switch can take any receiver out of the system if it blocks the voting.

Calculating the theoretical improvement in talk-back capability. The theoretical improvement in talk-back probability with voting receivers can be calculated using the following formula:

$$P = [100\% - (100\% - P_B)(100\% - P_{S1})(100\% - P_{S2})...(100\% - P_{Sn})]$$

where

P =   probability of a mobile unit being received when a voting system is used and where the base receiver is one of the voting receivers

P_B =   probability of a mobile being received when only the base receiver is used

P_{Sn} =   probability of a mobile being received by voting receiver S_n

Assume that a mobile or portable is positioned so that the base receiver receives it only 50% of the time; that is, $P_B = 50\%$. Six voting receivers are placed in a circle around the base station, as shown in figure 6-2. Assume P_{S1} is 80%, P_{S2} is 70%, and P_{S3} is 40%. The other receivers do not pick up enough signal to vote. Substituting values for P_B, P_{S1}, P_{S2}, and P_{S3} in the formula above gives us

$$
\begin{aligned}
P &= 100\% - (100\% - 50\%)(100\% - 80\%)(100\% - 70\%)(100\% - 40\%) \\
&= 100\% - (50\%)(20\%)(30\%)(60\%) \\
&= 100[1.00 - (0.50)(0.20)(0.30)(0.60)] \\
&= 100(1.00 - 0.018) \\
&= 100(0.982) = 98.2\%
\end{aligned}
$$

That is, the reliability of the talk-back capability has been improved from 50% to 98.2% by adding the voting receivers.

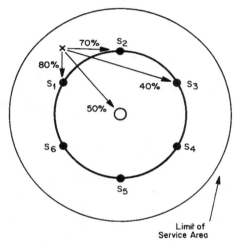

$\times$ = Mobile

● = Voting Receiver

○ = Base Receiver

Fig. 6-2 Improvement of Talk-Back Capability with Voting Receivers

The analysis above is theoretical. In practice voting receivers also are placed wherever there are dead spots or sites with poor reception. In addition the distribution of voting receivers depends on the availability of sites with ac power.

If hand-held portables are used, the number of voting receivers must be increased considerably because hand-held portables' lower power and antenna efficiency limit their range.

Each voting receiver's antenna height should be a compromise between maximum coverage and minimum interference. For example, a voting receiver in a large land mobile radio system was placed on the highest hill in the area. The voting receiver's antenna was

450 ft above sea level. It provided excellent coverage of the mobiles in the area. However, it also picked up an interfering signal of over 10 µV from a transmitter 75 miles away, which was enough to destroy communications.

Types of receiver voting systems. There are two general types of receiver voting systems used in voice land mobile systems. One is the discrete signal-level system where the "best" signal is the strongest signal. The other voting system is the status tone system where the "best" signal is the one with the best signal-to-noise ratio.

Discrete Signal-Level Voting System. The discrete signal-level voting system is an older method still in use in some land mobile radio systems. Audio tones are used to set up three discrete levels of voting. A low output signal causes the receiver to generate an audio tone. An output signal 6 dB higher generates a different tone or tones. When the receiver's audio output is yet another 6 dB higher or more, it generates a different tone. The tones are sent by telephone tie lines or microwave link to the comparator, where a tone decoder changes them to voltage levels. The comparator then selects the highest voltage level. The system does not discriminate between signal and noise, so the "best" signal is the strongest signal, even if the signal is pure noise.

An example of this voting system occurred in a power blackout in a large city. Voting receivers have batteries for operating when the power fails. One of the receiver batteries ran down after eight hours. This disabled the squelch circuit and resulted in a strong noise output. The comparator voted for the noise and the system locked up. No mobile unit could send a message until the offending receiver was rejected.

Status Tone System. A status tone is an audio tone the voting receiver sends to the comparator as long as the receiver is squelched. The status tone has two functions. If the comparator does not receive either a status tone or an audio signal, a warning light or alarm indicates a problem. The trouble could be with either the receiver or the telephone tie line. The status tone's other function is to maintain the comparator's line level by an automatic gain control (AGC) circuit. The status tone adjusts the AGC so that the received level at the comparator does not change.

The audio amplitude of all receivers is set at the same level at the comparator. A separate circuit in the comparator produces a dc voltage proportional to the noise from the receiver. The receiver with the smallest dc voltage (least noise) is the one voted. This is equivalent to the best signal-to-noise ratio.

Different manufacturers use different status tones, which range from 1950 to 2375 Hz. In some comparators jumpers can be used to accommodate any of the status tones.

It is extremely important that the amplitude of the audio level at the comparator for all of the voting receivers in a system is adjusted to the same level. This level is stated in the instruction books for particular voting systems. If some of the voting receivers are not set to the same level as the others, improper voting takes place. The voted audio does not necessarily have the best signal-to-noise ratio and could actually be a poor signal.

This situation occurred in a large emergency medical service radio system. The line levels of the different voting receivers at the comparator were not set at the same level. As a result the voting system decreased the ambulances' capability to talk back to hospitals. (Note: If a telephone tie line becomes noisy, a status tone voting system does note vote it.)

Voting systems and remote mobile relays.   A mobile relay consists of a receiver and transmitter, as described previously. When a carrier unsquelches a receiver, a squelch relay in the receiver squelch circuit turns on the transmitter. The receiver's audio output feeds the transmitter, which allows one mobile to communicate with another mobile through the mobile relay.

In the case of a voting system a switch circuit in the comparator turns on the transmitter by sending it the proper control tones. In addition the voted audio from the comparator is sent to the transmitter.

6.2 USING TRANSMITTER STEERING

Receiver voting systems are used frequently. Transmitter steering systems, illustrated in figure 6-3, are used much less. Transmitter steering works with a receiver voting system. All of the voting receivers from different sites are fed into a comparator. The voted audio from one site turns on only the transmitter at that site. Mobiles in the area of one site can talk to each other through the transmitter at that site. This way one pair of frequencies— one for the receivers and one for the transmitters—can cover a large area. However, not all radio systems are designed to add transmitter steering to a voting system. And note that transmitter steering does not cover the entire service area simultaneously—that need calls for simulcasting.

6.3 SIMULCASTING

In simulcasting systems audio is broadcast simultaneously over a number of transmitters on a single radio frequency. When two transmitters are on the same frequency, there is an overlap area. Within this overlap area is a noncapture area where the two RF signals produce an audible beat frequency in a mobile receiver. Differences in the phase and amplitude of the audio of transmitters cause distortion.

Minimizing Distortion

A high degree of the carrier frequency stability, audio amplitude equalization, and audio phase delay equalization minimizes the distortion.

Stability of the carrier frequency.   The beat note can be reduced either by synchronizing the frequencies or setting them to some specific frequency difference above or below the receiver's audio passband. These require high-stability oscillators with much greater precision than the conventional base-station transmitters. An example of a 450-MHz transmitter carrier frequency stability for a simulcast system is a specification of $\pm 0.000005\%$. If one transmitter is off $+0.000005\%$ and the other off -0.000005%, the two

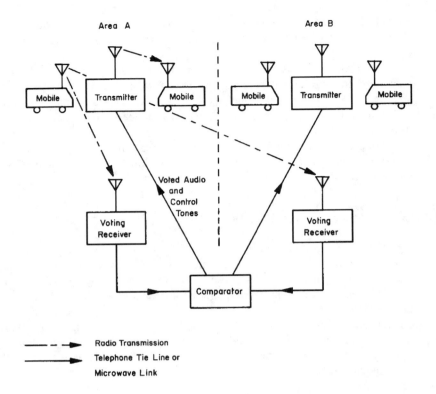

Area A   Area B

Voted Audio and Control Tones

Radio Transmission

Telephone Tie Line or Microwave Link

Fig. 6-3 Transmitter Steering

transmitters operating simultaneously would cause a maximum beat of 45 Hz. This measurement is well below the audio passband of 300 to 3000 Hz for land mobile radio systems, and therefore it would not be heard.

Equalizing audio amplitude. To avoid distortion the recovered audio amplitude for all transmitters and their control links should be the same within very close tolerances. Telephone tie lines are not recommended for simulcasting because of their short- and long-term variations in amplitude. Microwave links are used as control links for transmitters. All base transmitters and microwave links should use the same equipment throughout the simulcast system.

Equalizing audio phase delay. It is critical to equalize the audio phase delay in a simulcasting system. Phase equalization allows in-phase audio to leave each transmitter as close to the same time as possible. Each transmitter should have a delay equalization network.

CTCSS in simulcast systems. The CTCSS tones, although below 300 Hz, must still meet the amplitude and phase requirements of regular audio. To do this, a master

CTCSS oscillator and distribution system are used. The master CTCSS oscillator at the control point supplies the tones to all of the simulcast transmitters. Amplifiers at each transmitter site adjust the tones' amplitude. The phase equalization for CTCSS is accomplished by a phase-shifting network.

System maintenance.   Maintenance in a simulcast system is required more frequently and more thoroughly than in conventional systems. It requires specially skilled personnel.

Specialized test equipment.   Test equipment includes a storage oscilloscope in the audio range, a different audio phase meter, audio phase delay networks, and a phasing and monitor receiver. All of this equipment can be installed in a test panel near the control point.

Transmitter control panel.   The transmitter control panel is usually located at the control point near the simulcast test panel. This panel provides the transmitter controls necessary for testing, optimizing, and maintaining a simulcast system. Individual controls for each transmitter allow for normal, channel disable, or continuous transmission. Controls here allow for disabling CTCSS tones for each transmitter.

Weighing advantages and disadvantages.   Simulcasting's advantages are total area coverage and greatly extended range. It can be used to cover an entire county with only a pair of frequencies. On the other hand system design is much more complicated. Also, maintenance is increased compared to conventional systems and audio quality may be reduced in noncapture areas. (See References for Gary Gray's more detailed description of a working simulcast system.)

Second generation simulcasting.   Digitized voice radio simulcasting systems eliminate several of simulcasting's disadvantages. Digitized voice radio system standards are being developed by APCO Project 25 described in Chapter 10.

The system design for the second generation simulcasting is considerably simpler. The equalization of audio amplitude and audio phase delay is much easier with a digitized voice simulcasting system. The digitized voice simulcasting system overcomes many of the first generation system's maintenance problems, too.

The Motorola Astro Simulcasting System is an example of a second generation. This eliminates special simulcast test equipment and reduces infrastructure equipment such as delay modules, audio equalization modules, and special distribution modules. All of the critical filters are software-controlled. The system eliminates phasing delay procedures to compensate for variance with time or temperature. The digitized voice feature permits increased site separation.

6.4 USING A DIGITAL COMPUTER-CONTROLLED WIDE-AREA SYSTEM

Figure 6-4 shows a simplified form of a digital computer-controlled wide-area system. In practice there are many more areas, fixed receivers, and transmitters. A leased telephone line or microwave feeds the receivers' digital output to a computer in a network processor. If the first message the computer examines has no bit errors as determined by the cyclic redundancy check (CRC) the computer selects that message. If there are one or more bit errors, the computer goes to the next receiver. This continues until the computer receives a message without bit errors. There is no voting on the basis of best signal as in an ordinary voice voting system.

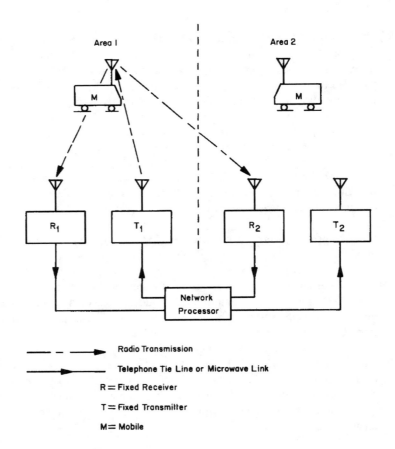

Fig. 6-4 Digital Computer-Controlled Wide-Area System

The computer activates the transmitter in the local area where the receiver has been selected. This way only one pair of frequencies can cover a large area. The system is not designed for voice transmission and works only for digital systems.

If the base transmitter initiates a communication and the mobile location is completely unknown, the computer can initiate a search algorithm that activates one transmitter after another until it finds the proper one for the specific mobile unit. Digital communication systems and CRC are examined in detail in Chapter 10.

REFERENCES

Cameron, Dennis W. "An Update on Simulcasting," *Mobile Radio Technology*. March, April, May, and June 1985.

Dewire, D. S., and E. A. Steere. "Precision Carrier Frequency Control and Modulation Phase Equalization of Base Transmitters in a Mobile Radio System," *IRE Transactions*. May 1960.

Gray, Gary D. "The Simulcasting Technique: An Approach to Total Area Coverage," *IEEE Transactions on Vehicular Technology*. May 1979.

Kaminski, W., and H. A. Schneider. "Multiple Base Transmitter Synchronization," *IEEE Transactions on Vehicular Technology*. October 1968.

7

Conserving Spectrum

At the present time the land mobile radio frequency spectrum is very crowded and additional frequencies are difficult to obtain. For this reason there has been a great deal of work in developing new techniques to conserve spectrum.

Trunking is one of these conservation systems. It takes advantage of the fact that some transmitters in a conventional system are idle at a particular time while others are busy.

Cellular radio systems use a number of small areas or cells for mobile telephones. Reusing frequencies over the entire covered area conserves frequency.

Amplitude-compandored single sideband (ACSB) produces a bandwidth of 5 kHz compared to a 30- or 15-kHz bandwidth in FM systems. Amplitude compandoring is used to obtain the low noise and other characteristics of an FM system. This chapter discusses all three techniques.

Historically, decreasing the bandwidth of FM transmitters has conserved spectrum. This chapter examines the limit of this type of reduction.

Digitized voice channels that use narrow 6.25 kHz bandwidths to allow for four new channels in a conventional 25 kHz bandwidth are discussed.

Spread-spectrum has been used mostly in government applications. The RF energy is spread at a low level over a wide spectrum. Some of the spread-spectrum techniques are described together with new applications in land mobile radio systems.

Different forms of multiplexing—CDMA, TDMA, FDMA, and so on—are described.

In 1992 the FCC proposed a refarming of the spectrum, which may include many of the techniques described in this chapter.

7.1 USING TRUNKING SYSTEMS

Trunking basically is a group of communications channels automatically sharing among a large group of users.

In a conventional nontrunked system mobile fleet, fleet 1 uses a dedicated communication channel, channel 1. Similarly, fleet 2 is assigned channel 2, fleet 3 has channel 3, and so on. If a vehicle from fleet 1 finds channel 1 busy, it is blocked even though channel 2 may be unused at the time.

In trunked systems the fleets are not assigned dedicated communications channels; the vehicle from fleet 1 is switched automatically to unused channel 2.

Comparing blocking times. When carriers are present on a conventional system 50% of the time (50% loading), a mobile user who wants access is blocked from the system 50% of the time. In a five-channel trunking system, with each channel loaded 50%, the blocking is only 12% of the time. In a 20-channel trunking, with each channel loaded 50%, there is practically no blocking.

Two trunking types. One method of trunking uses a dedicated data command channel as a control channel. All units not engaged in communications on other channels monitor the control channel. This means that four channels are available for communications in a five-channel trunking system. A dedicated data command channel is not used in the other trunking method. The control data information is sent out on the frequency that is the next available channel.

Trunking System with a Dedicated Data Command Channel. Motorola uses a dedicated data command type of system. All mobile and dispatcher control stations in the system are capable of operating on all channels in the trunked group. Each is equipped with a microprocessor that governs all mobile operations, including frequency, receive, and transmit selection.

A microprocessor called the system central controller governs all of the base-station repeaters at the base site. One of the radio channels is a data-only command channel that is used for call setup purposes only. The system central controller receives and processes service requests over this data-only command channel. All mobile and control stations not actively engaged in a voice message monitor the continuous stream of outbound data on the command channel.

The system's basic operation pattern: Dispatcher A wants to contact all of the units in fleet 1. He presses his PTT button, which causes the control station transmitter to send a short spurt of digital data. These digital data identify the calling unit and enter a voice channel request to the system central controller. The dispatcher's station then reverts to the receive mode to await a data response from the controller. On receipt of the request the system central controller selects a clear voice channel and sends out a data message over the data command channel directing all units in fleet 1 to the voice channel selected. Only units in fleet 1 automatically switch to this voice channel. The fleets that are not engaged in communications remain on the data command channel.

Trunking System without a Dedicated Data Command Channel.   At publication time, the E. F. Johnson Co.'s trunking system does not use a dedicated data command control system when fully loaded. Instead, it uses one of the repeaters in the system. The units assigned to the home channel use their home frequency for communication unless it is busy. In that case subaudible digital signals that share the voice channels switch a user's fleet to any available channel. If all radio channels are in use and a PTT button is pushed in a mobile unit, the receiver emits a busy tone.

Each repeater contains a logic module responsible for signaling on its own channel. The logic module in each repeater is connected to all other logic modules by a coaxial data bus. This way each repeater's logic module shares information.

Updating is accomplished by having an idle repeater send a 150-ms data burst every 10 seconds to its home-channel mobiles and control stations. Data transmission is continuous during communications on the home channel, with information to the mobile and control station logic units indicating which other channels are unoccupied. The data from the repeaters constantly switches the mobile and control units' idle repeaters to available open channels.

When a mobile operator pushes a PTT button, a data burst identifying the unit address is transmitted to the repeater. The repeater receives the data burst from the mobile. The repeater then transmits handshake data for the transmitting mobile. All other units monitoring the channel receive data that switch them to an available channel. After a transmission is completed, a turn-off code is transmitted, returning mobiles to the home channels.

Trunking system with a dedicated data command channel and failsoft.   In G.E.'s Enhanced Digital Access Communications System (EDACS) trunking system, a control channel failure results in a shift of most of its functions to another base station. Trunking cards located in each base station then control the trunking. This is called failsoft.

Siting and antennas.   Trunking equipment, base stations, and base-station antennas are all located at one premier site, usually a high building or tower. In a 20-frequency trunking system there are theoretically 20 receiver antennas and 20 transmitter antennas all mounted on one antenna tower. In practice a receiver multicoupler usually combines the 20 receivers into one antenna. The 20 transmitters also are combined by a transmitter combiner into another antenna mounted on the same tower. In some cases they also are combined into one antenna by using a duplexer. This enables a single antenna on top of a tower to have the best position for the entire trunking system. The duplexer, transmitter antenna combiner, and receiver multicoupler are discussed in Chapter 4.

Trunking systems and SMR.   Trunking systems are often operated by SMRs: entrepreneurs who own a trunking system and rent mobiles to people who are eligible to be licensed under Part 90 of the FCC Regulations.

Trunking and public safety radio service.   APCO instituted a project called Project 16 to develop functional guidelines for a public-safety trunking system. In 1978

APCO developed the specifications for a digitally addressed trunk radio system (DATRS). The final report, APCO Project 16A defines both mandatory and desirable performance capabilities for public-safety trunking systems. It can be purchased from APCO.

Problems with first generation trunking systems. Trunking systems using analog transmission may be referred to as first generation trunking systems. One problem with these systems is the lack of an effective encryption system to prevent unauthorized personnel intercepting information. This is a problem in public safety communications. Another difficulty is that one vendor's mobile unit cannot be used with another vendor's repeaters, so interoperability is lost between adjacent trunking systems of different vendors. This can be a severe problem in public safety communications between different jurisdictions.

Second generation trunking systems. Trunking systems where all mobile units use digitized voice are referred to here as second generation. Aegis digitized-voice mobile units that operate with EDACS trunking systems are one example. Digitized voice for mobile units allows sophisticated encryption, which is especially valuable in public safety communications. Another advantage of digitized-voice trunking systems: The received audio is of good quality throughout the service area, in contrast to analog-voice trunking where the voice quality deteriorates gradually as the mobile unit's distance increases. This advantage is due to the digitized-voice radio transmission's superiority and is not unique to digitized-voice trunking.

Future trunking standards. To establish digital trunking standards and solve problems such as the lack of interoperability, a trunking standards Task-Group was established as part of APCO Project 25, New Technology Standards Project, Digital Radio Technical Standards. At publication time the Task-Group on trunking is working on the development of standards for digital trunking. An interim report of APCO Project 25 including a report of the Trunking Task-Group was adopted January 15, 1993. It is expected the trunking standards will take some time before completion.

7.2 CELLULAR MOBILE RADIO SYSTEMS

A cellular mobile radio system is basically a system to reuse a number of frequencies over and over again. A service area is divided into small cells, each with a low-powered base station. The cells are grouped into larger units called clusters. Each cell within a cluster is assigned a different frequency. Adjacent clusters repeat the frequencies, as illustrated in figure 7-1, where 16 cells make up a cluster. Each number represents a different frequency of a low-powered base station within the cell. As a mobile goes from one cell to another, the mobile frequency is changed to correspond to the base frequency of the new cell. In practice each cell has a set of frequencies so that a number of vehicles in a cell can make a call simultaneously.

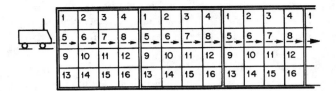

Fig. 7-1 Reusing Frequencies in Cellular Radio Systems

Switching mobile frequencies. A mobile telecommunications switching office (MTSO) switches frequencies by using a hand-off routing. As the signal level decreases in one cell and increases in another, the mobile telecommunications switching office signals the mobile to change frequency. Another cell site takes the hand-off and continues the phone calls without interruption as the mobile goes from one cell site to another.

Cell splitting. Initially seven hexagonal cells may be used as a start up system. When the number of vehicles increases as the system matures, the cells can be split. One method is to split each of the seven cells into three sectors by using three 120° directional antennas. This results in 21 cells per cluster.

Frequency assignments. In one commercial application each cell in a cluster of 21 cells is assigned a different set of 16 channels. This allows a number of vehicles in a cell to make a call to the base station simultaneously. A frequency management chart lists the specific 16 frequencies assigned to each of the 21 cells. It is used to minimize co-channel interference. An example of a frequency management chart can be found in the article, "Elements of Cellular Mobile Radio Systems," listed in the References.

Two carriers per service area. The FCC assigns two companies to each service area. Each carrier is assigned one-half of the available spectrum. One company is a wire-line carrier and the other is a nonwire-line carrier.

Co-channel interference. Co-channel interference is an important consideration in a cellular system. The ratio of signal to interference, S/I, should be at least 18 dB. One factor in co-channel interference is D/R, the co-channel reuse ratio. D is the separation between the two co-channel cells and R is the radius of the cells. In a hexagonal-shaped cellular system $D/R = \sqrt{3N}$, where N is the number of cells in a cluster. In addition directional base station antennas in a cell can be used to reduce co-channel interference between cells.

Advanced mobile phone service (AMPS) operation

Figure 7-2 shows the communication paths of the advanced mobile phone service developed by Bell Laboratories. The control center is the MTSO. There is one MTSO for each service area.

The mobile unit contains a combination radio transceiver and logic unit in the trunk. This is connected to the antenna and the mobile phone control unit near the driver.

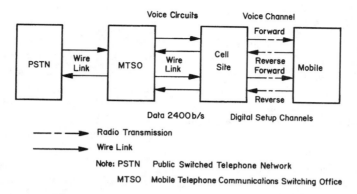

Fig. 7-2 Communication Paths in AMPS

At each cell site there is a designated setup channel where digital information is transmitted continuously between the cell site and the mobile unit. When the mobile unit is turned on, it selects and monitors a particular setup channel. The mobile unit listens for a paging signal containing the mobile phone number in binary form.

When the call originates from a fixed station, the person at the fixed station dials the mobile unit's number. This goes through the public switched telephone network (PSTN) to the mobile's home MTSO in a specific service area. The MTSO converts the mobile's phone number to the mobile's identification number. The MTSO then instructs the cell sites that have paging channels to page the mobile over the forward setup channel. This paging signal is broadcast over the entire service area. Because the mobile unit observes its setup channel continually, it detects its number and then seizes control of the reverse setup channel. As soon as it does so, the mobile unit transmits its identification number to the cell site. The cell site in turns sends a message to the MTSO by way of a 2400-bit/second (b/s) wire data link. This message tells the MTSO that the called unit has responded. The MTSO assigns an idle voice channel to the mobile unit by sending a message through the cell site and then over the forward setup radio channel to the mobile unit. The mobile unit moves to the assigned voice frequency. The MTSO then directs the cell site to transmit a data message over the radio voice channel. This data message rings the bell in the mobile unit. The mobile unit answers, the MTSO removes the audible ringing circuit, and the conversation starts.

When the call originates from a mobile unit, the mobile subscriber dials the desired phone number while on-hook. This permits resending without redialing if the called number is busy. It also allows for the mobile subscriber to correct errors before going off-hook. The mobile subscriber then depresses the "send" button on the handset unit, which causes the mobile unit to transmit a digital message over the reverse setup channel to a nearby cell site. The digital message contains the unit's identification number, the called number, and a request for a voice channel. The cell site sends this to the MTSO over the wire data link. The MTSO in turn, through the cell site, tells the mobile unit which of the cell site's voice channels to use for the call. After the mobile unit is on the correct voice channel, the MTSO connects the call through the PSTN. When the called party picks up the phone, the connection is made.

If the mobile unit terminates the call first and goes on-hook, the mobile unit transmits a signaling tone to the cell site, which then places an on-hook signal on the land-line trunk that was used in this communication. The MTSO in turn sends disconnect signals through the PSTN. The MTSO also shuts down the appropriate cell's radio transmitter.

When the land party terminates the call first and goes on-hook, the mobile unit receives the disconnect signal via the PSTN. The MTSO then sends a transmitting message via the cell site to the mobile unit. The mobile unit proceeds with the same course as when the mobile unit terminates the conversation.

One of the most important functions in AMPS is the hand-off as the mobile goes from one cell to another. The hand-off is the transfer of the mobile unit from one radio channel to another. The MTSO transfers the call to an idle radio voice channel at an adjacent cell site. At the same time the cell site blanks the voice signal for about 50 ms. During the voice signal blanking, a burst of data on the voice channel instructs the mobile unit to switch to the new channel. The entire hand-off function takes about 200 ms, which allows the call to continue without interruption.

The cellular base station regularly sends a supervisory audio tone (SAT) of about 6000 Hz to the mobile unit. The mobile unit retransmits the SAT back to the cellular base station. If the cellular base station does not receive the SAT, it is assumed the mobile unit has ceased transmitting.

Cellular disaster operations.   Some cellular companies have begun to distribute special cellular phone packages to emergency communication personnel during disasters such as floods and hurricanes. Priorities must be designated in advance so non-emergency communications are cut off automatically during disasters.

Cellular call boxes.   The installation of fixed cellular phones on highways to replace emergency call boxes presents another use for cellular radio. Maintaining the underground wires of conventional telephone call boxes on highways is difficult and expensive. Cellular Communications Inc., in Lake Forest, California, has installed over 1000 fixed cellular phones on freeways in Orange County. The cellular phones are mounted on special breakaway poles. Solar panels and chargeable batteries supply the power for the units. The cellular phones will be in direct communications with the California Highway Patrol headquarters. Similar equipment has been installed in many freeways and turnpikes in the United States.

Second generation digital cellular.   Digital cellular systems have begun operating all over the world. They have increased the cellular system's capacity by a factor of 6 per carrier, improving the system efficiency and conserving spectrum. Digital cellular is described in more detail in Chapter 10.

7.3 SQUEEZING IN MORE CHANNELS WITH AMPLITUDE-COMPANDORED SINGLE SIDEBAND

In a conventional single-sideband system amplitude modulation is used instead of frequency or phase modulation. In amplitude modulation the radio-frequency carrier has an upper sideband and a lower sideband. For example, an RF signal of 100 MHz is modulated with an audio tone of 1000 Hz (or 0.001 MHz). The resulting radio frequency consists of the RF carrier of 100 MHz, an upper sideband of (100 + 0.001) MHz, and a lower sideband of (100–0.001) MHz. In single-sideband operation one sideband is eliminated and the carrier is suppressed: if it is the lower sideband, the radio frequency is the upper sideband, or 100.001 MHz. The RF carrier of 100 MHz is reinserted at the receiver. The maximum bandwidth of the single-sideband transmitter is approximately 3 kHz.

To their advantage single-sideband systems take up much less spectrum space than conventional FM transmitters. For example, in the 72 to 76-MHz band, transmitters are separated by 30 kHz. As many as six single-sideband transmitters can be substituted for the one FM transmitter.

Overcoming Single-Sideband Disadvantages

Conventional single-sideband systems have three main disadvantages when compared to FM systems. The first disadvantage is noise since amplitude modulation picks up all sorts of electrical machinery and other radio noise. Also, when the single-sideband receiver re-inserts the carrier frequency, it may not fall precisely in the right place, which may cause problems in a message's readability. Thirdly, conventional FM systems exhibit a capture effect and AM systems do not. In the capture effect if there are two FM base carriers on the same frequency, only the stronger signal is heard, provided it is at least 6 dB (four times the power) more than the weaker signal.

In amplitude-compandored single-sideband (ACSB) two basic improvements developed by Bruce Lusignan of Stanford University minimize the disadvantages. Pilot tone and compandoring make the difference.

Pilot tone.   The pilot tone is the insertion of a fixed precise tone at 3.1 kHz above the suppressed carrier at the transmitter. This takes care of ordinary single-sideband's inaccurate tuning when the carrier frequency is reinserted in the receiver. The receiver also has a pilot-tone generator. A phase detector in the receiver compares the transmitter pilot tone with the receiver pilot tone. This results in a dc voltage that controls the receiver local oscillator to bring it in synchronization with the transmitter.

Compandoring.   Compandoring is essentially compressing the audio signal at the transmitter and expanding it in the receiver. This eliminates two ordinary single-sideband's disadvantages when compared to FM: noise and the lack of the capture effect. The dynamic range of ordinary speech is about 30 dB. In compandoring compression is used in the transmitter to reduce the dynamic range to about 7.5 dB. In the receiver the dynamic audio's range is expanded back to 30 dB. Figure 7-3 shows the audio signal's compression and expansion and its effect on noise. The diagram shows that the noise in the receiver in the compressed state is only about 7 dB down from the peak audio signal. The expansion amplifier is nonlinear; it amplifies large inputs more than small inputs. After expansion in the receiver the noise is down almost 30 dB from the audio peak. This reduction in noise levels tends to cancel out FM's advantage in reducing the effects of noise. In addition another signal, some 7 dB below the desired signal, is reduced to almost 30 dB below the desired audio after expansion. This result is similar to the capture effect in FM.

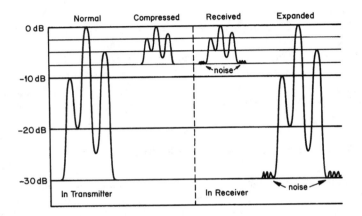

Fig. 7-3 Compandoring in ACSB

Using ACSB to conserve spectrum.   Figure 7-4 shows the interleaving of one ACSB channel between two existing adjacent FM channels. Each FM channel has a 5-kHz guard band, which results in 10 kHz between the two FM stations. This allows for an ACSB transmitter, which occupies 3.1 kHz between the two FM transmitters.

Figure 7-5 shows another method of using ACSB to conserve spectrum. Six ACSB transmitters are substituted for one 30-kHz FM channel in the 72- to 76-MHz band.

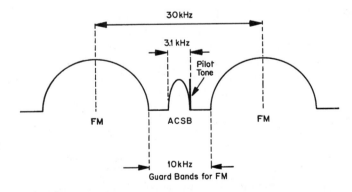

Fig. 7-4 Interleaving One ACSB Channel between Two FM Channels

Chap. 7   Conserving Spectrum

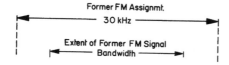

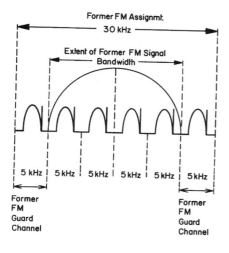

Fig. 7-5 Substituting Six ACSB Channels for One FM Channel in 72- to 76-MHz Band

7.4 REACHING THE LIMITS OF NARROWBAND FM

Land mobile radio's history has been one of narrowing the frequency separation between adjacent channels. In the 72- to 76-MHz band the separation was reduced from 60 to 30 kHz. The bandwidth was also reduced from 60 to 30 kHz in the 150–174 MHz but with geographical separation between adjacent channels 15 kHz apart. In the UHF bands (421–430, 450–470, and 470–512 MHz), the original separation was 100 kHz, which was reduced to 50 kHz and then to 25 kHz in 1967. In 1970 special low-power 12.5-kHz offset channels were introduced.

There are limitations to this procedure. However, it may still be possible to reduce the separation between adjacent channels even further. In Japan and in some parts of Europe the separation has been reduced to 12.5 kHz. By reducing the maximum deviation from the present 5 kHz to 2.5 kHz, it may be possible to use 12.5-kHz separation between channels in the United States and conserve spectrum even further. The FCC has proposed to use 12.5-kHz separation or even lower in some of the newer 800- and 900-kHz frequencies.

According to the National Bureau of Standards Technical Note 103 FM systems are superior to AM systems if the modulation index is more than 0.6 and the rate of improvement is proportional to the square of the modulation index. The modulation index for an FM system is defined as the ratio of the maximum deviation of the RF carrier to the maximum significant frequency of a modulating pure sine wave. This puts a limit to narrowband FM.

7.5 NARROWBAND DIGITIZED VOICE CHANNELS

With new technology it is now possible to digitize voice, transmit it in a narrowband, and then covert it back to voice at the receiver end. There is a program, APCO Project 25, that is preparing standards for narrowband digitized voice. APCO Project 25 proposes first taking a 25-kHz channel and splitting it into two digitized voice channels, each 12.5 kHz. Then at some future date, each 12.5-kHz digitized voice channel may be split into two 6.25 digitized voice channels. The technology is described in Chapter 10.

7.6 SPREAD-SPECTRUM TECHNIQUES

The FCC set up a UHF Task Force in 1976 to study methods of increasing land mobile radio's spectrum efficiency. One of the systems the FCC studied then and is still pursuing is spread-spectrum.

Spread-spectrum uses a wide bandwidth to transmit signal information. The energy is spread across many conventional channels but at a much lower level than conventional systems. An ordinary narrowband receiver would not detect the spread-spectrum signal.

There are a number of different spread-spectrum technologies, but they all have some common characteristics: The transmission of information covers a bandwidth much wider than would be normally required in a conventional radio system, and the transmission is coded by a pseudo-random sequence that both the transmitter and receiver share. In a pseudo-random sequence the coding appears to be random but is actually repeated cyclically. Different pseudo-random sequences are assigned to distinguish different users.

Specific Spread-Spectrum Techniques

Frequency hopping.   In frequency hopping a wide band of frequencies covering several megahertz is divided into a large number of much narrower channels. The transmitter hops from one channel to the other, transmitting very short bursts. The hopping sequence is pseudo-random generated according to a key that is available to the transmitter and its intended receiver but to no other receiver. This system was originally developed by the military to prevent jamming by the enemy.

Direct sequence.   At the transmitter each bit in the original message is multiplied by the same pseudonoise (PN) sequence. A PN is a rapid cyclical pseudo-random transmission of + and -1^s. It has a wide spectrum similar to white noise. Each original bit after multiplication is divided into pieces called chips. A chip corresponds to a member of a PN sequence. The chip is transmitted as a +1 or a −1 depending on whether its corresponding element in the PN sequence is a +1 or −1. For example, if the original bit is +1 and the PN sequence is +1, +1, +1, −1, −1, +1, −1, then after multiplication, the chip sequence would be +1, +1, +1, −1, −1, +1, −1. If the original bit is −1, then the chip sequence is −1, −1, −1, +1, +1, −1, +1. The spectrum of the resulting bit stream is spread out over a wide bandwidth.

At the receiver a special filter, whose values correspond to the PN sequence at the transmitter, obtains the original information bit stream. Only the receiver with the proper filter components can detect the original message.

Direct sequence was originally developed by the military to obtain secure voice communications. The FCC has proposed allowing police departments to use direct sequence spread-spectrum for secure communications.

Time hopping.   In time hopping the RF carrier is pulsed and its period and duty cycle are varied under the direction of a pseudo-random sequence. This is sometimes combined with frequency hopping.

Applying spread-spectrum.   In the past most applications of spread-spectrum technology have been in the military. However, the technology now is used in some portable telephones and in some cellular digital systems. A form of spread-spectrum has been proposed for use in the SMR service. Spread-spectrum is also used in personal communication networks.

7.7 MULTIPLEXING TO INCREASE SPECTRUM EFFICIENCY

Multiplexing is the combining of multiple circuits in a single transmission over a single channel.

Code-division multiple access.   CDMA is a form of multiple access spread-spectrum systems use. It is based on the principle that each user is distinguished from all others by having a unique code that is imprinted on the transmission in one of several ways. The receiver uses the code to select the proper transmission from among the many spread-spectrum circuits in operation. CDMA is used in some digital cellular radio systems and in some personal communication networks.

Time-division multiple access.   In the TDMA system each channel is assigned a time slot. In the United States the basic TDMA system combines 24 circuits for transmission, with each circuit having its own time slot. This is the basic T-1 carrier, with the multiplexed channel operating at 24 times the data rate of each channel. The format is called Digital Signaling 1 (DS-1) and the facility is referred to as a T-1 carrier. Higher speed allows multiple T-1 groups to combine into T-2 carrier, T-3 carrier, and so on. There is also a corresponding combination of formats, DS-2, DS-3, DS-4, combining DS-1 formats. Section 8.2 gives a detailed description of TDMA system operation. Some digital cellular radio systems and many control consoles operate with TDMA.

Frequency hopping multiple access. FHMA technology uses a large number of narrow channels as described in frequency hopping, Section 7.6. It is sometimes used with TDMA. The FCC has proposed FHMA for use in the SMR. A special code causes the transmitter to hop from one frequency to another in a pseudo-random manner. The same code in the receiver causes it to hop from one frequency to another in the same way.

Frequency-division multiple access. There are a number of relatively narrow frequency channels in FDMA. FDMA may be either analog as illustrated in Section 8.2 or digital.

7.8 REFARMING THE SPECTRUM

On November 6, 1992 the FCC[*] released a notice of rule making that includes refarming or improving the spectrum efficiency standards of the Private Land Mobile Radio Service frequency bands below 512 MHz. The proposal includes gradually splitting each 25-kHz channel in the 421- to 430-, 450- to 470- and 470- to 512-MHz bands into four 6.25-kHz channels. Gradually splitting each channel in the 150- to 174-MHz band into three new 5-kHz channels is included. Each 30 kHz channel in the 72- to 76-MHz band is to be gradually split into six 5-kHz channels, also.

The FCC did not specify the technology to be used. However, digitized voice can obtain the 6.25-kHz bandwidth, as described in Sections 7.5 and 10.12. Similarly, ACSB can obtain 5-kHz channels, as described in Section 7.3.

Proposed time scale for the 421- to 512-MHz bands. The FCC proposed that the original 25-kHz channel's frequency deviation be reduced from 5 kHz to 3 kHz by January 1, 1996. This would create a pseudo 12.5-kHz channel flanked on either side by a 6.25-kHz channel. The next step is to convert the pseudo 12.5-kHz to a true 12.5-kHz digitized voice channel. The FCC proposed beginning four-channel conversion starting in 2004. The central 12.5-kHz channel will be split into two 6.25 kHz digitized voice channels, which will result in a total of four 6.25-kHz digitized voice channels. This conversion is to be completed between 2008 and 2012. Radio systems in congested cities are to be finished first.

Proposed time scale for the 150- to 174-MHz band. Existing users are to reduce frequency deviation to reduce occupied bandwidth to 12 kHz by January 1, 1996. This would permit the FCC to eliminate adjacent channel mileage separations. Starting in 2004 three 5-kHz new channels would be created for each original channel. This conversion is to be completed between 2008 and 2012. Radio systems in congested cities are to be finished first.

---

[*] FCC Notice of Proposed Rule Making, Docket No. 92-235. Replacing Part 90 with Part 88.

Chap. 7   Conserving Spectrum

Proposed time scale for the 72- to 76-MHz band.   Existing users in the 72- to 76-MHz band are to reduce occupied bandwidth to 10 kHz by January 1, 1996. Thus three channels would be created from every existing channel. A 10-kHz channel would be centered on the original channel's center frequency and be licensed to all existing users. The other two channels would be 5 kHz wide, spaced just above and below the 10 kHz, and would be available for new users. All users in the 72- to 76-MHz band are to employ 5-kHz channels starting in 2004. The conversion is to be completed between 2008 and 2012. Radio systems in congested cities are to be finished first.

Other FCC refarming proposals.   The FCC has proposed including 800-MHz band to increase spectrum efficiency. The FCC also proposes using direct sequence spread-spectrum for public safety covert operations. Another proposal is to decrease the antenna's power and height to improve spectrum efficiency.

Strategy to maintain a system during refarming.   The rapid change of radio equipment that some of the FCC refarming programs dictate requires special planning; it is difficult to buy a new, large system at one time because of budget and other restrictions. One strategy is to buy backwards compatibility transmitters and receivers that can operate in the old mode but, thanks to a switch, can operate in the required new mode. For example: A Motorola Astro Transceiver in the 421- to 512-MHz band operates in the conventional analog mode with a 25-kHz bandwidth. However, the equipment comes with a built-in switch. Activating the switch causes the transceiver to operate in a voice digitized mode with 12.5-kHz bandwidth. This type of backwards compatibility transceiver gradually can replace all of the conventional analog equipment. Then the switches can be activated and the system can operate with digitized voice in a 12.5-kHz bandwidth, which would meet the FCC refarming proposal in the 421- to 512-MHz band. However, backwards compatibility equipment does not exist for the proposed FCC refarming in the 72- to 76- and 150- to 174-MHz bands which call for 5 kHz. At the time of this writing there is a great deal of opposition to the proposed 5 kHz channels in the 72–76 and 150–174 MHz bands. The backwards compatibility strategy depends on the final FCC timetable for refarming.

REFERENCES

Barghaussen, A. F., et al. National Bureau of Standards Technical Note 103, 1965. Available at a depository library for Government publications, such as Columbia University Engineering Library.

Childs, Fred B. "CSSB-ACSB, The Progression of Narrow Band VHF/UHF Technology," *Proceedings of the Radio Club of America, Inc.* Vol. 58, No. 1. Spring 1984.

Cooper, G. R., and R. W. Nettleton. "A Spread-Spectrum Technique for High-Capacity Mobile Communications," *IEEE Transactions on Vehicular Technology.* November 1978.

Dixon, Robert C. *Spread Spectrum Systems.* 2nd ed. New York: John Wiley & Sons, Inc., 1984.

Hirshman, P. "Companding Land Mobile Radio SSB Modulation to Increase Number of Channels in VHF and UHF Bands," *Spectrum.* 1979.

Kavanagh, Donald. "APCO Project 16A, 900 MHz Trunked Communications, System Functional Requirements Development". 1979. This can be obtained from APCO, Suite 202, 2040 South Ridgewood Avenue, South Daytona, FL 32119.

Lee, William C. Y. *Mobile Cellular Telecommunications Systems.* New York: McGraw-Hill Book Company, 1988.

_____ . "Elements of Cellular Mobile Radio Systems," *IEEE Transactions on Vehicular Technology.* May 1986.

Reeves, Charles M. "An Overview of Trunking Techniques in Mobile Radio Systems," *Proceedings of IEEE-VTS Convergence '80 Conference.* IEEE CH 1601–4. Utica, NY: Bohn Printing Co., 1980.

8

Ties That Bind

Many land mobile radio systems contain remote base stations, control points, voting receivers, and so on, that are separated by distances of a few miles. To connect these components different types of tie lines are in use. These include primarily dedicated telephone tie lines and microwave links, although other methods are also used, including fiber optics, closed-circuit TV, and radio-frequency links.

The history of tie lines has been the use of telephone tie lines first and then the gradual substitution of microwave links. This has increased lately as the monthly cost of telephone lines has gone up sharply. This cost rise has prompted some municipalities to look into still other methods, such as using planned closed-circuit TV systems.

Tie lines are an essential part of many large land mobile radio systems. They also can be a major source of trouble as well as expense. It is with this in mind that the various types of tie lines are discussed in this chapter, starting with telephone tie lines.

8.1 USING TELEPHONE TIE LINES

It is important to know the types of telephone tie lines, the expected losses in the lines, the maximum amount of signal the telephone company allows into the lines, the frequency response, and the circuit noise.

Types of Telephone Tie Lines

There are three basic types of telephone tie lines most frequently used in mobile radio systems. First and oldest are the dc or metallic lines, which are copper lines from one piece of radio equipment to another. Either dc or tones can be used with metallic tie lines. Today these lines are difficult to obtain from a telephone company, but there are older radio systems that still use them. The second and most common are the carrier voice-grade lines, called 2002 and 3002. Direct current cannot be used on these lines; instead, tones in the 300- to 3000-Hz range are transmitted. The third type of line is a conditioned line for transmitting data.

Metallic telephone lines.   Metallic lines are used to control base transmitters by applying dc current. Metallic lines also are used in receiver voting systems.

Maximum Signal Power.   The American Telephone and Telegraph Company has published a document, *Bell System Transmission Engineering Technical Reference, PUB 43401*, that lists its requirements for metallic circuits. From Figure 2 of this document, the single-frequency-tone maximum permitted signal for a balanced 600-Ω line is:

At 1 kHz: approximately +10 dBm
At 2 kHz: approximately –2 kBm
At 3 kHz: approximately –6 dBm

Maximum Voltage and Current.   The restrictions on operating voltage and current for dc lines, from *PUB 43401*: Magnitude for peak voltage between any conductor and ground must be less than 70.7 V. The exceptions are for both continuous and interrupted dc voltage, where the magnitude, or peak, of dc voltage between any conductor and ground must be less than 135 V. The recommended maximum current is 150 mA per conductor.

Signal Losses, dB/Mile.   The losses in balanced dc lines depend on the wire gauge, the distance, and the leakage resistance to ground. The wire size of copper telephone lines varies from American wire gauge (AWG) 19 to AWG 26. The attenuation in dB/mile at 1000 Hz is 1.3 for AWG 19, 1.7 for AWG 22, 2.1 for AWG 24, and 2.7 for AWG 26.

Repeaters.   The telephone company may add repeaters to metallic circuits to compensate for lines losses when a line is very long (over 12 miles). The repeaters can be used with dc, are bidirectional, and have a gain that can be adjusted up to 12 dB.

Leakage Resistance to Ground.   The leakage resistance of the wire to ground is usually more than 100,000 Ω for metallic circuits.

Metallic Circuit Noise.   Metallic circuit noise at the interface is usually less than 20 dBrnc, where O dBrn is equivalent to –90 dBm and the c refers to readings made with a C-message[*] weighting network.

Carrier-type lines.   Carrier-type lines do not have dc continuity and tones must be used instead of dc current levels for remote control of radio transmitters.

Equalization of Lines.   The frequency response of carrier telephone tie-line circuits in the 300- to 3000-Hz band is attenuated at the lower and higher portions of the band.

---

* See Appendix I.

To compensate for this attenuation some radio systems include equalization equipment for use with telephone tie lines. Equalizing the lines enables the amplitude of the frequencies from 300 to 3000 Hz to be approximately the same at the far ends of the telephone tie line. This improves the speech intelligibility as well as compensates for the telephone tie line drop-off at certain tones at the high end of the band. These tones are used in voting systems and in controlling remote transmitters.

Types of Carrier Lines for Radio Systems. Two general types of carrier lines can be used for radio systems: 2002 and 3002. Only the 3002-type lines are recommended by radio engineers. The specifications for the 3002-type lines are given in *Bell System Technical Reference PUB 41004*, published by American Telephone and Telegraph Company. This publication gives the specifications for both the basic 3002 lines and the conditioned 3002 lines.

Basic 3002 Lines Specifications. Here is a summary of specifications common to standard design voice-band data channels and important for radio tie lines, from *PUB 41004*:

Maximum design loss at 1004 Hz is 16 ± 1 dB + 4 dB long-term variation, or a total loss of approximately 20 dB.

Maximum losses at 2175 and 1950 Hz, unconditioned: 20 dB + 8 dB = 28 dB.

Maximum noise 28 dBrnc (–62 dBm) for lines up to 50 facility miles measured with Western Electric Co. type 3 noise meter. A measurement with an AC-VTVM should give a reading of –50 dBm or less for an acceptable line for mobile radio systems.

Nominal impedance is 600 Ω balanced.

Voice band is 300 to 3000 Hz.

Maximum input power for status tones is 0 dBm.

Maximum input power in general: 0 dBm averaged over 3 seconds. Instantaneous power limit is +13 dBm.

Maximum frequency shift is ±5 Hz. Carrier systems operate in a single-sideband suppressed carrier mode. Since the carrier must be reinserted in this system, there is a difference in frequency between modulating and demodulating carriers. This frequency shift can degrade some data demodulation processes.

Conditioned 3002 lines. Conditioned lines are used primarily for data transmission. These lines improve the frequency response of telephone lines and reduce envelope distortion.

Envelope distortion is the delay of some frequency components more than others; that is, a signal's frequency components travel at different speeds. Among other reasons, the delay at low frequencies is caused by the inductive effects of transformers in the telephone system, while capacitive effects cause the delay at high frequencies. The filters in telephone carrier systems also contribute to envelope distortion.

Envelope distortion does not affect the ear in speech transmission. Digital signals, however, can be so distorted by envelope distortion that they may not be recognizable. A nonflat frequency response from the telephone lines can distort digital signals, also. Ideally a telephone transmission channel would have sharp cutoff frequencies at each end of the channel. In actual practice the frequency response rolls off at each end of the band. This can be caused by the low-pass characteristics of loaded cable or by the high-pass characteristics of transformers and series capacitors.

There are five degrees of line conditioning listed in *PUB 41004* (p. 13). Each degree provides tighter specifications on envelope distortion and frequency response. These lines are referred to as basic (unconditioned), C-1, C-2, C-3, C-4, and C-5. C-1 and C-2 lines are used in two-way radio systems for transmitting data.

PUB 41004's frequency response and envelope distortion are listed in table 8-1 for the basic (unconditioned), C-1, and C-2 lines.

TABLE 8-1 BASIC AND CONDITIONED 3002 LINES

Type of Channel Conditioning	Envelope Distortion		Frequency Response	
	Frequency Range (Hz)	Variation (μs)	Frequency Range (Hz)	Variation* (dB)
Basic (uncondi-tioned)	800–2600	1750	500–2500	−2 to +8
			300–3000	−3 to +12
C-1	1000–2400	1000	1000–2400	−1 to +3
	800–2600	1750	300–2700	−2 to +6
			300–3000	−3 to +12
C-2	1000–2600	500	500–2800	−1 to +3
	600–2600	1500	300–3000	−2 to +6
	500–2800	3000		

* (+) means loss with respect to 1004 Hz; (-) means gain with respect to 1004 Hz.

8.2 SUBSTITUTING MICROWAVES FOR TELEPHONE TIE LINES

Microwaves are used in many land mobile radio systems. They can replace telephone tie lines in receiver voting systems and handle remote control of base stations. They also send large amounts of information simultaneously from one point in the mobile system to another. In some cases microwaves are used as part of a private telephone system for city or county communications.

Microwaves in Voting Systems

This section emphasizes the use of microwaves as a substitute for telephone tie lines in voting systems and in the control of remote transmitters.

Applicable FCC rules. The FCC Rules and Regulations, Part 94, governs the private operational fixed microwave service. Under these rules anyone eligible for licensing under FCC Parts 81 (Stations on Land in the Maritime Services), 87 (Aviation Services), and 90 (Private Land Mobile Services) is eligible for microwave operation under Part 94.

Operational microwave bands. Part 94 authorizes frequencies within the following bands (listed in megahertz): 900; 1800; 1900; 2000; 6000; 10,000; 12,000; 13,000; 17,000; 18,000; 19,000; 20,000; and 30,000.

Multiplexing. Microwaves can send a large amount of information by multiplexing. Multiplexing is the simultaneous transmission of a number of voice or data channels on one radio-frequency carrier. Figure 8-1 shows microwave duplexing for receiver voting in a land mobile radio system. There are two methods: FDMA and TDMA.

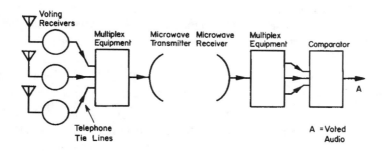

Fig. 8-1 Microwave Multiplexing in Land Mobile Radio Systems

FDMA. FDMA is one widely used method of multiplexing. In FDMA the total frequency spectrum available is divided into channels, each of which occupies a particular frequency range all the time. The bandwidth for a single voice channel is 4 kHz. The various channels' frequency allocations usually follow the CCITT modulation plan.

In simplified form the single-sideband suppressed carrier of a HF (high-frequency) subcarrier performs the multiplexing in a channel modem for each voice channel. Each modem's outputs are then combined into what is known as the baseband. The baseband frequency modulates the RF microwave transmitter. There is different equipment available with different numbers of channels. Figure 8-2 shows a baseband of 60 to 2540 kHz with 600 CCITT channels. In practice orderwire and alarms are added, as shown in figure 8-2. Orderwire refers to a voice channel at each microwave station used for intercommunications during maintenance. Alarms are the signals transmitted to an attended station indicating failures or malfunctions, such as high temperatures, ac failure, low dc voltage, equipment failure, and excessive receiver noise.

Groups and Supergroups in FDMA. A CCITT standard group consists of 12 channels, each 4000 Hz wide. Five standard groups are combined into a supergroup. Ten supergroups are combined to form 600 channels in the baseband.

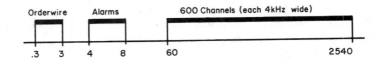

Fig. 8-2 Frequency-Division Multiplexing (Frequency in Kilohertz)

Using Data with FDMA. Digital transmission over a 4-kHz voice band can be sent up to 9.6 kb/s (9600 bits per second).

Channel Drop. The backbone of a microwave system is the routing between the two principal terminal sites. It may contain a number of intermediate repeaters between these two terminals. In a channel drop, one or more channels in a multiplex system are terminated at some point intermediate between the end terminals of the system.

MUX Drop. MUX (multiplex) drop is the same as channel drop. It may refer to all of the multiplex channels being terminated at some point intermediate between the end terminals of the system. New channels then may be inserted at this point for further transmission.

Noise Load Ratio (NLR). NLR is the ratio of the noise power applied to the entire baseband to the nominal signal power appearing in a single channel.

Noise Power Ratio (NPR). Noise power ratio describes intermodulation noise performance. There is a certain amount of intermodulation noise with an FDMA system's separate channels due to the frequency interaction of various channels. NPR is the ratio in decibels of N_C to N_T. N_C is the noise in a test channel when all channels are loaded with white noise. N_T is the noise in a test channel when all channels except the test channel are loaded with white noise. White noise is noise with a flat spectrum at the frequency range of interest.

Direct to Line. "Direct to line" means the direct application of a baseband to microwaves. A microwave system that is not direct to line: The baseband modulates an intermediate frequency (IF) of 70 MHz, and then the modulated IF is up-converted to a microwave frequency.

TDMA. Most microwave transmissions in land mobile radio systems use FDMA. However as telephone costs rise, more local governments are considering microwaves as a local bypass compatible with AT&T-Bell Telephone carriers. These carriers use TDMA in both land wire and microwave systems.

The first step in TDMA is to sample each channel in succession. For a 4-kHz channel, the minimum sampling rate is 8000 samples per second. Each channel has a gate that enables its output only when the gate is open. If we assume a 24-channel system, the gate is open for a very short time for each channel in turn. The full sequence of the 24 channels' sampling is called a frame. There are 8000 frames per second since this is the required minimum sampling rate for a 4-kHz channel. All of the sampled 24 channels form a pulse amplitude modulation (PAM) wave.

The next step is to change the PAM into a digital-coded stream of bits through pulse-coded modulation (PCM). A quantitizer divides the PAM into 128 amplitude levels. Each level of signal then is converted into a seven-digit code such as 1 0 1 1 1 0 1. Physically the digit "1" is a rectangular pulse and the digit "0" is represented by the absence of a pulse. The 1 and the 0 are referred to as bits. The stream of bits modulates the microwave carrier. The AT&T-Bell companies use this system in a digital carrier called T1. To calculate the number of bits in a T1 carrier:

The T1 carrier has 24 PCM voice channels with a seven-digit code. In addition to these seven bits, there is one signaling bit and one framing bit. The total number of bits per frame is

$$(7 + 1 \text{ signaling bit}) \times (24 \text{ channels}) + 1 \text{ framing bit} = 193 \frac{\text{bits}}{\text{frame}}$$

Since there are 8000 frames per second, the total bit rate is

$$193 \frac{\text{bits}}{\text{frame}} \times \frac{8000 \text{ frames}}{\text{second}} = 1{,}544{,}000 \text{ bits per second}$$

This is usually expressed as 1.544 mb/s. This can be compared to the TDMA system, which can handle only 9600 b/s of data. A number of T1 carriers can be time-multiplexed in a microwave carrier. There are also T2 carriers and other combinations of T1 carriers.

TDMA has its hierarchy of groups, supergroups, and so on, just as FDMA does. The basic group for TDMA in the AT&T-Bell system is the DS1 or digigroup. The digigroup transmits 1.544 Mb/s and is used to time multiplex 24 digitally encoded voice-frequency circuits. Four DS1 signals can be combined into a DS2 superdigigroup with 6.312 Mb/s. Seven DS2 superdigigroups can be combined into a DS3 masterdigigroup with 44.716 Mb/s. Six DS3 masterdigigroups form a DS4 signal with 274.176 Mb/s.

Line-of-Sight Microwave Transmissions

Having discussed multiplexing in microwaves, here are the problems of obtaining line-of-sight microwave transmission:

Factors in obtaining line of sight. At microwave frequencies propagation is mainly line of sight. However, there are a number of factors to be considered in obtaining line-of-sight transmission. These include earth bulge, refraction, and diffraction.

Earth Bulge. Since the earth is curved, there is a bulge between two microwave towers, as figure 8-3 shows. The amount of earth bulge in feet at any point in a path is $h = 0.667d_1d_2$ for an unbent radio ray. Here h is the earth bulge at any point in feet, d_1 is the distance in miles from the near end of the microwave link to the bulge, and d_2 is the distance in miles from the earth bulge to the far end of the link.

Refraction of Radio Waves. Radio waves are refracted, or bent, as they go from denser air levels to thinner air levels, or vice versa, similar to the bending of light waves as

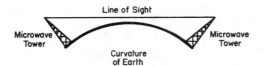

Fig. 8-3 Earth Bulge

they enter or leave water. Normally the refraction bends away from the earth, making the earth's bulge less. Sometimes the refraction causes inverse bending, which increases the earth's bulge. Mathematically this modifies the formula for earth bulge as follows:

$$h = \frac{0.667 d_1 d_2}{K} \quad \text{where } K = \frac{\text{effective earth radius}}{\text{true earth radius}}$$

As figure 8-4 shows when K is greater than 1, the ray is bent away from the earth, which allows users to lower the towers. If K is less than 1, the ray is bent toward the earth and the towers would have to be higher. For most planning purposes K is taken to be 4/3.

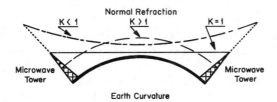

Fig. 8-4 Microwave Paths as a Function of K

Fresnel Phenomenon or Diffraction. Fresnel phenomenon is an additional amount that must be added to clear an obstacle. A wave front expands as it travels, resulting in reflections and phase changes as it passes over an obstacle. This in turn causes an increase or decrease in the signal level received. To avoid the effect an additional clearance is added to the obstacle. The Fresnel phenomenon occurs in zones. The accepted added clearance is 0.6 of the first Fresnel zone radius in feet, or

$$\text{added clearance} = 0.6 \, (2280) \left[\frac{d_1 d_2}{L \, (d_1 + d_2)} \right]^{1/2}$$

where

d_1 = distance from transmitter to path obstacle in statute miles

d_2 = distance from receiver to path obstacle in statute miles

L = wavelength of microwaves in ft

Combining the factors for line of sight. In practice, charts and graphs for Fresnel and other calculations are used to obtain a clear line of sight for a microwave system. A profile chart showing all obstacles between two proposed radar towers is drawn up on rectangular graph paper. The obstacles' heights above sea level are taken from topographic maps. Midearth bulge is considered an obstacle and should be computed and marked. Then the earth curvature and the Fresnel zone clearance are added to each obstacle height at that point. Allowance also is made for tree growth (50 ft) and for existing vegetation (10 ft). When all the obstacle points are plotted, the minimum tower heights can be determined by drawing a straight line between the tower sites, passing just over the highest obstacle point.

Microwave equipment. The microwave equipment basically consists of a transmitter, antenna transmission lines, antennas, and receivers.

Typical Microwave Transmitter. Figure 8-5 illustrates a typical microwave transmitter used in land mobile radio systems. Other configurations are used also.

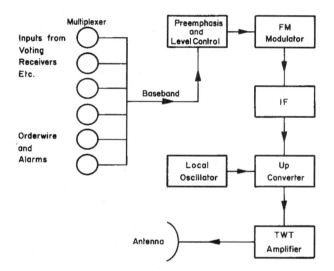

Fig. 8-5 Typical Microwave Transmitter

Multiplexer. The multiplexer already has been described in some detail. The various inputs to the microwave system are combined in the multiplexer to form the baseband. This is accomplished by each input modulating a separate carrier in the single-sideband suppressed carrier (SSBSC) mode to form 4-kHz slots in the baseband.

Preemphasis. In preemphasis the higher baseband frequencies' deviation is increased. That is, the amplitude of the higher baseband frequencies is increased relative to the other baseband frequencies. This is done because the thermal noise of the FM receiver increases with increasing baseband frequency, which means a decreasing signal-to-noise ratio with increasing baseband frequency. To compensate, the signal at the higher baseband frequencies in the transmitter is increased relatively so the signal-to-noise ratio in the receiver is uniform across the baseband.

FM Modulator and IF Stage. The baseband frequency modulates an IF stage.

Up-Converter. The modulated IF is mixed with a local oscillator in an up-converter stage to form the final modulated microwave carrier. In a mixer the two inputs can be subtracted or added. In an up-converter the two input frequencies are added.

Traveling Wave Tube (TWT). A TWT is used as a power amplifier at microwave frequencies. It is an electron tube with an electron gun, a long electron beam surrounded by a helix, and input and output cavities. The helix propagates a slow wave and the electron beam interacts with the wave on the helix. The result is the amplification of the wave when the electron stream to the helix gives up dc energy as radio-frequency energy.

Antenna Transmission Lines. Below 2 GHz, coaxial cables are used as antenna transmission lines from the microwave transmitter to the antenna. From 2 GHz, the antenna transmission lines are rectangular, elliptical, or circular waveguides.

At one time microwave installations used only rectangular waveguides. However at the present time, to avoid losses in bends with the rectangular waveguide, elliptical waveguides are used in new installations where a single length is required from microwave equipment to the antenna. Circular waveguides are used to minimize losses for long runs. The waveguide discussed here is a rectangular waveguide. The long side of a section of a waveguide is one-half wavelength or more, as figure 8-6 shows.

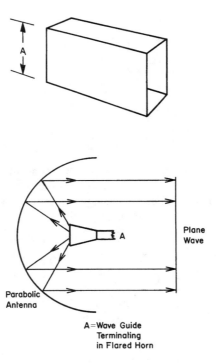

Fig. 8-6 Section of a Rectangular Waveguide (Note: Dimension A must be one-half wavelength or more)

Fig. 8-7 Parabolic Antenna

At low frequencies where the waveguide's long dimension is less than one-half wavelength, the attenuation is very high. A one-half wavelength is considered the cutoff frequency. The waveguide should not be operated below frequencies about twice cutoff. The latter is because of interference higher-order propagation modes set up. A flared horn may match the waveguide to radiate into free space. The flared horn can be directed into a metallic parabolic antenna as shown in figure 8-7. The waveguide itself is always maintained under pressure by using dry air or nitrogen to prevent moisture within the waveguide.

Antennas.   The parabolic antenna reradiates the energy in a narrow beam of high gain. In microwaves an antenna's gain is usually given as so many dB above an isotropic antenna. An isotropic antenna is a theoretical antenna that radiates equally in all directions. This is sometimes called effective radiated power (ERP), which is the sum of the transmitter's power output plus the antenna's gain (referenced to a dipole antenna) minus antenna transmission-line losses. The effective isotropic radiated power (EIRP) is the same as ERP except that the antenna's gain is referenced to an isotropic antenna. A parabolic microwave antenna's gain over an isotropic is given in decibels as

$$G = 20 \log D + 20 \log F + 7.5$$

where

D is in feet, F is in gigahertz, and the antenna efficiency is assumed to be 55%.

Typical Microwave Receiver.   Figure 8-8 shows a typical microwave link receiver in land mobile radio systems.

Antenna and Transmission Line.   These components are identical to the transmitter.

Local Oscillator and Down-Converter.   The antenna transmission-line output is mixed with the local oscillator in the down-converter. This results in an IF output for the down-converter.

Limiter and FM Discriminator.   These parts are the same as in an ordinary FM receiver.

Noise Figure of Receiver.   The amount of noise the receiver introduces is expressed as the ratio in decibels of the noise at the receiver's output to the noise at the input.

Noise Threshold of Receiver.   The noise threshold of a receiver at room temperature in decibels below 1 W is

$$N_{TH} = -204 \text{ dBW} + 60 \log(\text{IF bandwidth}) + \text{noise figure}$$

where

N_{TH}	=	noise threshold
$-$dBW	=	dB below 1 W
IF bandwidth	=	bandwidth, MHz

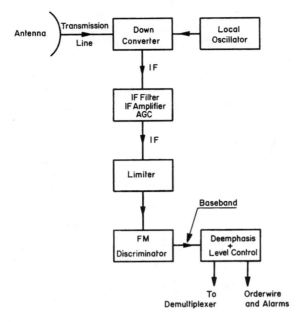

Fig. 8-8 Typical Microwave Receiver

Say a receiver has an IF bandwidth of 10 MHz and a noise figure of 10. Then the noise threshold = −204 dBW + 60 log(10) + 10, or the noise threshold = −134 dBW.

Carrier-to-Noise Ratio. The difference in decibels between the received signal level and the receiver's noise threshold level is the carrier-to-noise ratio. For example, if a signal into a receiver is −80 dBW and the noise threshold of the receiver was −134 dBW, the carrier-to-noise ratio is −134 − (−80) = 54 dB. If the signal into the receiver is exactly −134 dBW, the carrier-to-noise ratio is 0.

Minimum Carrier-to-Noise Ratio (C/N). For many calculations the minimum *C/N* is 10 dB, to use FM's "capture effect." In the case of a receiver with −134 dBW, the minimum *C/N* is −134 dBW + 10 dB = −124 dB. This is the minimum usable signal.

Fade Margin. Fade margin is an additional decibel amount built into the system design to compensate for fading. It is directly tied in with the design path reliability figure. Fade margin is the difference in decibels between the actual signal and the minimum usable signal.

Free-space propagation. In microwaves free-space propagation usually is used for calculations. To understand the formula for free-space loss consider an isotropic antenna radiating equally in all directions. The free-space path loss in decibels is

$$\text{free-space path loss} = 36.6 + 20 \log D + 20 \log f$$

where D is the distance between the transmitter and receiver antennas in statute miles and f is the frequency in megahertz.

Simplified line path gains and losses. Assume a 1-W transmitter (0 dBW), a transmitter and receiver antenna gain of 25 dB each (referenced to an isotropic antenna), and line losses at each end of 2.5 dB. The transmitter antenna is 10 statute miles from the receiver antenna. There is a line of sight between the two antennas. The frequency for this hypothetical example is 6 GHz or 6000 MHz. The first step is to calculate the free-space path loss from the previous formula. In decibels:

$$\text{the free-space loss} = 36.6 + 20 \log 10 + 20 \log 6000$$
$$= 132.16 \text{ dB}$$

Add all the gains and losses as follows:

Transmitter output	0.00 dBW
Transmitter line losses	−2.50
Transmitter antenna gain	+25.00
Free-space path loss	−132.16
Receiver antenna gain	+25.00
Receiver line losses	−2.50
	−87.16 dBW

Assume the receiver has a noise threshold of −134 dBW; the minimum usable signal is −134 dBW + 10 dB = −124 dBW. The difference between the actual signal and the minimum signal is −124 + 87.16 = −36.84 dB. This is the fade margin for the system, which compensates for fading and improves path reliability.

Fading. Fading is a temporary random decrease in the microwave transmission's signal level. There are two types: multipath and power fading. The interference between a direct wave and a reflected wave from an atmospheric layer causes multipath fading. Multipath fading may be more than 30 dB for seconds or minutes. Power fading is due to the temporary partial blocking of the transmitter-receiver path. One cause is precipitation, which may produce a fading of 20 to 30 dB for hours.

Path reliability. Path reliability is the percentage of time that a specified signal-to-noise ratio is met. A path reliability of 99.999% means the performance criteria is not met for 0.001% of the time. In a year this is equivalent to an outage time of 5.26 minutes. Incorporating a fade margin into the system design compensates for outage time due to fading. There are two diversity methods of improving path reliability:

Space Diversity. Space diversity systems use a second antenna spaced several wavelengths from the original antenna in the vertical plane. Using these two antennas at different heights takes advantage of the fact that multipath fading does not occur at the same time at each antenna. A switching arrangement chooses the antenna that is receiving the greatest signal.

Frequency Diversity. Frequency diversity systems take advantage of the fact that maximum fades do not occur at the same time at different frequencies. In a frequency diversity system two transmitters and two receivers paired to two different frequencies are in continuous operation. If the signal at one frequency tends to fade, the other frequency still comes through at normal levels. A signal comparator in the receiving system selects the greater incoming signal. However, frequency diversity is difficult to implement because the additional frequency may not be available in many metropolitan areas. Also, there is the expense of additional transmitters and receivers.

Equipment-system reliability. Equipment-system reliability is increased by using a hot standby.

Hot Standby. A hot standby is a parallel microwave equipment that comes on line automatically when the primary microwave equipment fails or malfunctions. For example, if a transmitter's frequency or power changes by a preset amount, the hot standby transmitter is switched automatically in its place. Similarly, if a receiver AGC voltage or squelch fails, a hot standby receiver is substituted automatically.

Repeaters. There are two basic types of repeaters: active and passive. An active repeater is an intermediate station in a microwave system that receives a signal from a distant station, amplifies it, and retransmits the signal to another distant station. This is usually performed in both directions simultaneously.

Active Repeaters. There are two basic types of active repeaters: the baseband and the heterodyne.

Baseband repeaters—With baseband repeaters the received signal is demodulated to the original baseband frequencies and reinjected onto the transmitter's modulator for retransmission. Each demodulation and modulation in a microwave system introduces noise. The additional noise may be troublesome in a multihop system with a number of repeaters. Using heterodyne repeaters may avoid this problem.

Heterodyne repeaters—In an IF heterodyne repeater the received signals are converted to an immediate frequency using a local oscillator and mixer. The resulting IF is then up-converted to a new radio frequency, which is amplified by a TWT and then retransmitted.

Passive Repeaters. Passive repeaters are devices (not antenna elements), such as a place reflector, that are put in the microwave transmission path to redirect the microwave energy. They can be used when one of the microwave antennas is on a small building. A plane reflector on a tall building reflects the signal to and from another microwave transmitter-receiver. The gain of the plane reflector in decibels above an isotropic antenna is

$$\text{gain} = 20\frac{4\pi\,A\,\cos a}{L^2}$$

where

L = wavelength, m
A = surface area, m^2
a = 1/2 horizontal angle between incident and reflected waves

Cost Analysis of Microwaves Versus Telephone Tie Lines[*]

As the cost of telephone tie lines increases sharply, other methods, such as microwaves, become more economically feasible. In general the more telephone tie lines to replace, the more probable a microwave replacement would be cheaper. A land mobile radio system designed for walkie-talkie radios has approximately eight times the number of voting receivers (and telephone tie lines) compared to a system designed primarily for vehicular radios. Thus there is a greater probability that microwaves are a more economical replacement for telephone tie lines in walkie-talkie radio systems than in vehicular radio systems.

Figure 8-9 shows the telephone tie lines and microwave system used in this comparison. Sixty telephone tie lines connect 60 voting receivers to a comparator.

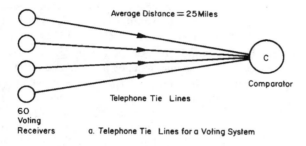

60 Voting Receivers   a. Telephone Tie Lines for a Voting System

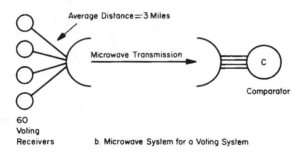

60 Voting Receivers   b. Microwave System for a Voting System

Fig. 8-9 Substituting Microwaves for Telephone Tie Lines

Cost of Telephone Tie Lines. [*] There are three basic charges for most telephone tie lines: a one-time initial connection charge, a monthly rental charge per mile, and a monthly circuit charge. These charges vary from area to area. However, for the sake of illustration, assume an initial connection charge of $250 per telephone tie line, a monthly rental of $30

---

[*] The cost figures used in this book are for demonstration only. The actual cost figures must be determined at the time the comparison is made.

per mile, and a monthly circuit charge of $20. In this example the average distance of the telephone tie line is 25 miles. The telephone system expenses may be separated into initial and annual costs.

Initial costs—For this example the initial connection charges are $250 × 60 tie lines or $15,000.

Annual costs—The annual rental is $30 per month per mile × 12 months × 25 miles × 60 tie lines, or $540,000. The annual rental by circuits is $20 per circuit per month × 60 × 12, or $14,000. The sum of the two annual charges totals $554,000.

Cost of Microwave System.*

The charges for the microwave system may be divided similarly into initial and annual costs.

Initial costs—The initial expenses consist of system design, site preparation, equipment and installation, and initial connection charges for supplementary telephone tie lines. The system design expenses include a frequency search and the development of a detailed technical specification. The cost of the system design for this example is $75,000. The site preparation costs include complete equipment room construction, estimated for this example at $150,000. The cost of the microwave equipment, including multiplexing for 60 channels, is presumably $300,000. The initial connection charge for the 60 short telephone tie lines from the voting receivers to the microwave transmitter is $15,000. The total initial costs of the microwave system are $75,000 + $150,000 + $300,000 + $15,000, or $540,000.

Annual costs—The annual charges include vendor maintenance, utilities, and the supplementary telephone tie lines' cost. Vendor maintenance is estimated at $30,000 per year for this example. The cost of utilities is presumably $1000 per year. The 60 supplementary telephone tie lines in this example average 3 miles from the voting receivers to the microwave transmitters. The annual rental by the mile is $30 per month per mile × 12 months × 3 miles × 60 lines, or $64,800. The annual charge per circuit is $14,400 for the 60 lines. The annual cost for the supplementary telephone lines is therefore $78,000. The total annual cost of the microwave system is $30,000 × $1,000 × $79,200, or $110,200.

Comparing telephone tie-line and microwave costs.

Compare the two costs by drawing two time-flow charts from time 0 to the end of 15 years (the estimated life of the microwave system). It is assumed that at the end of the 15 years the microwave system will have negligible salvage value.

Telephone Tie-Line Costs (Thousands of Dollars)							
15	554.4	554.4	554.4	554.4	554.5		554.4
0	1	2	3	4	5	...	15
			Years				

Microwave Costs (Thousands of Dollars)							
540	110.2	110.2	110.2	110.2	110.2		110.2
0	1	2	3	4	5	...	15
			Years				

* The cost figures used in this book are for demonstration only. The actual cost figures must be determined at the time the comparison is made.

The two systems may be compared directly by converting all costs to present worth. Present worth is the cost of a system at time 0 on the time-flow chart. The calculations take into account the cost of money, which is assumed to be 10% compounded annually in this example. Present worth can be calculated as follows:

$$P = R\left[\frac{(1+i)^{n}-1}{i(1+i)^{n}}\right]$$

where

P	=	present worth
R	=	uniform series end-of-period payment
i	=	interest rate at end of each period
n	=	number of periods

The term in brackets is called the uniform series present worth factor (USPWF). It can be evaluated by using a calculator or referring to the tables in *Managerial and Engineering Economy* by George Taylor, listed in the References. For this example, where i = 10% and n = 15 years, the USPWF is 7.6061.

Telephone Tie Lines' Present Worth.   Since the annual cost of the telephone tie lines is $554,000,

$$P = 554,400(7.6061) = \$4,216,821.84$$

The initial cost of the telephone tie lines, $15,000, must be added to this. Thus the telephone tie lines' present worth is $15,000 × 4,216,821.84, or $4,231,821.84.

The Microwave System's Present Worth.   Since the annual cost of the microwave system is $110,200,

$$P = 110,200(7.6061) = \$838,192.22$$

The initial cost of the microwave system, $540,000, must be added to this. Thus the microwave system's present worth is $540,000 × 838,192.22, or $1,378,192.22.

Comparing the Two Systems' Present Worth.   The present worth of the microwave system is $1,378,192.22, compared to $4,231,821.84 for the telephone tie lines. The microwave system is therefore the economical choice in this example. The present worth analysis can be used to compare the costs of different types of tie lines in land radio mobile systems.

The Break-Even Point.   The break-even point is the number of years, n, it would take to get back the extra investment in the microwave system. The value n is evaluated as follows:

initial cost A + n(annual cost A) = initial cost B + n(annual cost B) where A is the telephone system and B is the microwave system. Substituting values yields

$$15,000 + n(554,400) = 540,000 + n(110,200)$$
$$n = 1.1 \text{ years}$$

8.3 SEEING THE LIGHT WITH FIBER OPTICS

Fiber optics are rapidly supplanting copper telephone wire. In land mobile radio systems fiber optics can be used to convey data more efficiently than copper wires. A 0.005-in.-diameter fiber can carry the same information as a 3-in. bundle of 900 pairs of copper cable. This translates into a much smaller space and a much lighter weight for optical fibers.

Basic Fiber Optics Communication System

Figure 8-10 illustrates the basic fiber optics communication system. The driver in the transmitter converts the input signal into a form that can modulate the light source. The light-emitting diode (LED) is the light source used for short distances (less than 5 miles) and data rates lower than 100 Mb/s. The more powerful injection laser diode is used for longer transmission and higher data rates. The optical fiber used to transmit the modulated light has a minimum attenuation at a wavelength of 1.55 microns,* which equals a frequency of 193,548 GHz.

History.   Actual wavelengths used for light transmission started out with 0.8 microns in first-generation systems. Second-generation systems were designed to operate at 1.3 microns. Today third-generation systems operate near 1.55 microns for minimum attenuation. The optical fiber connects the transmitter to the receiver. In the receiver a light detector converts the modulated light to an electrical signal, which is then fed to the output circuit. Photodiodes are used as light detectors.

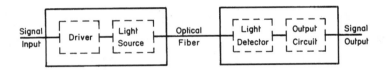

Fig. 8-10 Basic Fiber Optics Communication System

Light Sources.   Both the LED and the injection laser diode are semiconductor chips. The LED is cheaper and is used for shorter distances and lower data rates.

LED—The LED basically operates like an ordinary semiconductor diode, but the LED gives off light energy instead of heat energy. This is due to the special materials used in the LED.

Injection laser diode—The basic laser (light amplification by stimulated emission of radiation) was patented in 1958. The laser diode used in light communications was first demonstrated in 1962. The laser emits coherent, monochromatic light, which is different from ordinary light. Coherent means that all the emitted light waves are in phase, rising and falling together. Monochromatic is the emission of one wavelength only. In practice the ba-

---

* A micron is a millionth of a meter.

sic injection laser is not completely monochromatic but has a narrow frequency range of light. The laser emits a narrow, intense beam of light that does not spread like the LED.

Optical Fibers. The optical fiber is made of two glass materials: the core and an outer layer called the cladding. An outside jacket covers the cladding. There are two general types of cable, depending on the method of light transmission:

Multimode optical fiber—In the multimode cable type the light travels both by total internal reflection and straight transmission. It gathers relatively large amounts of light and can be used with an LED for short distances. As Figure 8-11a shows, light that goes straight down the core arrives at the far end first. The other rays the core-cladding interface reflects arrive at the far end later. This spreading of light is called modal dispersion and it can cause an increase in pulse rise and fall times. The multimode fiber is the least expensive and the easiest to work with and terminate. Its disadvantages, including modal dispersion and low efficiency, limit its use to short runs and lower operating frequencies.

Single-mode optical fiber—The core diameter is reduced in the single-mode optical fiber. The fiber transmits only one mode, as Figure 8-11b shows. It is expensive and requires a laser to provide the necessary precise source of light. The single-mode optical fiber is used for high-speed long-distance applications.

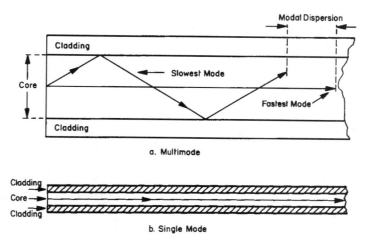

Fig. 8-11 Light Transmission Optical Fibers

Light Detectors. Several devices are used to convert the photons of light to an electric current. Two of these are the positive intrinsic network (PIN) photodiode and the avalanche photo diode (APD).

PIN—The PIN has a positively doped area, a small undoped, or intrinsic, area, and a negatively doped area. Without light current cannot flow across the intrinsic section, which acts like a large depletion area. One photon of light striking the intrinsic area produces one electron hole pair, which flows through the diode. There is no amplification in this device.

APD—In the APD device each photon of light results in a number of electrons flowing through the diode. It is more sensitive and faster than the PIN, but it is more expensive.

8.4 USING CABLE TELEVISION IN A TWO-WAY RADIO SYSTEM

The rising cost of leased telephone lines, used to connect voting receivers among other functions, has forced land mobile radio system operators to look at other methods. One such system, microwaves, has been described. Another method is to take advantage of a planned cable television system in a city. Many cities have contracts with cable television (CATV) companies that give the municipality free access to the CATV facilities. This makes using the cable system financially feasible for city-operated land mobile radio systems.

Basic CATV System

CATV consists of coaxial cables with amplifiers spaced at intervals along the cable to overcome attenuation.

Cable configurations. Conventional TV entertainment cable installations use the subsplit cable system. The band from 50 to 300 MHz goes in the forward direction, while 5 to 30 MHz goes in the reverse direction. Video conferences, telephone, and data transmission use the midsplit cable system. The band from 5 to 120 MHz goes in the forward direction, while 170 to 440 MHz goes in the reverse direction. The dual-cable system consists of two separate cables, each with its separate amplifiers. One cable is used for forward and the other for reverse transmission.

Cable. At present CATV uses polyethylene dielectric coaxial cable with an aluminum sheath. The size varies from 0.25 in. for subscriber house drop applications to 1.0 in. for long trunk applications. In the future fiber optics may supplant the coaxial cable in new installations.

Amplifiers. To compensate for the signal's attenuation, amplifiers are inserted at different locations to amplify the signals back to their original level. The amplifiers' gain varies from 20 to 35 dB. The cable supplies power for the amplifiers. Battery standby power for CATV amplifiers is a requirement if the system is used as an interconnection for two-way radio systems.

Terminal equipment. The terminal equipment includes the modulator and demodulator. The modulator or transmitter that feeds the cable usually has an output between 1 to 10 mW. CATV engineers use the notation of dBmv or dB referenced to 1 mV in their work. 0 dBmv corresponds to 1 mV across 75Ω. This is the approximate input level that produces a good TV picture for the individual subscriber. In terms of dBmv the CATV transmitter has an output level of +50 to +60 dBmv. The demodulator or receiver has a much wider bandwidth and lower sensitivity compared to communication receivers.

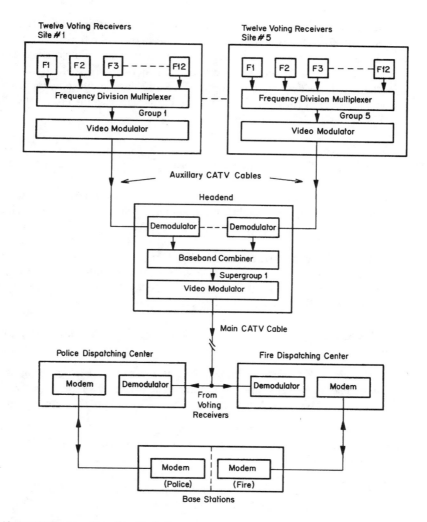

Fig. 8-12 CATV Used in a Two-Way Radio System

Connecting a two-way radio system into CATV. Figure 8-12 shows a hypothetical radio system with 12 frequencies and five sites for voting receivers. Each of these sites has 12 receivers, one for each frequency. At each of the voting sites there is a frequency-division multiplexer (FDM), which takes the 12 audio outputs to form a group. The group then frequency modulates a video modulator that is an RF transmitter with a bandwidth of about 6 MHz. The coaxial cable sends the five voting sites' multiplex outputs to the headend, the terminal equipment of the CATV system. The output of each voting receiver site is demodulated at the headend into its five separate groups, which are then combined to form supergroup 1. The supergroup frequency modulates a video modulator and the output is transmitted over the main CATV cable to the police and fire dispatching centers. Suitable filters and demodulators separate out the audio for each of the two dispatching centers. At each of the dispatching centers modulator-demodulators (modems) send audio and control tones to the repeater. Here the cable is two-way. Although this radio system is hypothetical the principles and equipment would be used with some modifications to fit the particular case. For further information on this subject, see the articles by Frank Genochio and Jay W. Underdown listed in the References.

8.5 LINKING THE SYSTEM WITH RF

As the cost of telephone tie lines goes up drastically, radio-frequency links to replace short telephone tie lines in a land mobile radio system become more practical. These RF links are low-power radio systems usually between 1 to 5 W.

UHF splinter frequencies for RF links. In the UHF-band frequencies offset by 12.5 kHz from standard frequency assignments are available for transmitters below 2W to use as RF links. FCC licensing limitations restrict the transmitter antenna height to 20 ft. The receiver antenna has no restrictions on height.

Types of RF links. RF links can be simplex or duplex, depending on the radio equipment.

Simplex RF Link. A simplex RF link is a transceiver operating on 110-V ac with a standby battery power supply. The unit can be used as a simplex two-way RF link. It can also act in a receive-only or transmit-only mode by deleting the appropriate crystal.

Duplex RF Link. The transceiver is connected to an antenna duplexer, which makes simultaneous transmit and receive operation possible. Separate transmit and receive frequencies must be used.

RF link uses. RF links are used to transmit voting receiver's output to a comparator. They also are used to control remote base stations.

Voting Receivers. At the voting receiver site a transmit-only RF unit, operating continuously, sends the voting receiver's audio output to a receive-only unit at the comparator. The RF link acts exactly as a telephone tie line would.

Base Station Control. Figure 8-13 shows the RF link system used to remotely control a base station. At the dispatcher console end is a simplex transceiver RF link with two-wire PTT audio. Tone control is used to key up the base transmitter. The base station to be controlled requires four-wire audio, tone control, and a squelch gate output to key the RF link. The base station end of the RF link is duplex, which allows the dispatcher to have priority over mobile calls. This particular system uses 1-W UHF transmitters operating on offset frequencies.

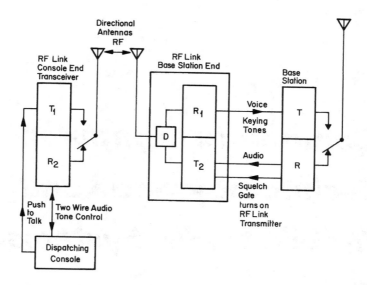

Fig. 8-13 Link for Base Station Control

REFERENCES

Cheung, W. Stephen, and Frederick Levien. *Microwaves Made Simple: Principles and Applications.* Dedham, MA: Artech House, Inc., 1985.

Genochio, Frank. "Some Uses of CATV Technology in Land Mobile Communications," *IEEE V.T.S. Conference Record.* 1981.

Harrell, Bobby. *The Cable Television Technical Handbook.* Dedham, MA: Artech House, Inc., 1985.

Taylor, George. *Managerial and Engineering Economy.* New York: Van Nostrand Reinhold Company, Inc., 1964.

Underdown, Jay W. "The Application of Cable Television in the Design of a Radio Communication System." Spectrum Resources, Inc., P. O. Box 1141, St. Charles, MO 63302, 1984.

White, Robert F. "Reliability in Microwave Communication System—Prediction and Practice." Lenkurt Electric Co., San Carlos, CA, 1970.

Yeh, Chai. *Handbook of Fiber Optics: Theory and Application.* San Diego: Academic Press, Inc., Harcourt Bruce Jovanovich, 1993.

Zanger, Cynthia and Henry. *Fiber Optics Communications and Other Applications.* New York: Merrill, Imprint of Macmillan Publishing Co., 1991.

Publications without listed authors

"Data Communications Using Voice Band Private Line Channels," *Bell System Transmission Engineering Technical Reference PUB 41004.* October 1973. Obtainable from AT&T Customer Information Center, Commercial Sales Representative, P. O. Box 19901, Indianapolis, IN 46219.

"Microwave Path Engineering Considerations," Lenkurt Electric Co., San Carlos, CA. September 1961.

9

Combining Computer Technology and Land Mobile Radio Systems

The computer, whether a microcomputer on a chip or a large computer, is being used more extensively in land mobile radio systems: frequency synthesis and selection in two-way radios, mobile data terminals, radio consoles control, trunking, cellular, and CAD. It is important to have some knowledge of microprocessors and microcomputers, at least in a general sense. While radio people whose education is recent may be conversant in computer technology, there are many who may not be that familiar with certain aspects regarding the control of radio systems and devices. This chapter introduces basic computer technology, especially microcomputers and microprocessors, then describes some of the microprocessors used in radio systems. Finally, this chapter looks at two computer technology uses in land mobile radio systems: console control and CAD.

9.1 REVIEWING BASIC APPLICABLE COMPUTER TECHNOLOGY

A microcomputer is a small digital computer that, like most digital computers, consists of four basic components—input, central processor, memory, and output—as figure 9-1 shows. Buses—groups of parallel conductors— tie the components together. A microcomputer can vary in size from one or two printed circuit boards to an entire unit on one chip. The central processor in a microcomputer is called a microprocessor.

Microprocessor word length. The microcomputer is a digital device. Data, instructions, and addresses within the microcomputer must be in binary form or bits. A specific number of bits in parallel transfer the data. For example, early microprocessors transmit eight bits at a time: Eight conductors in the data bus, one for each bit, transfer the data information around the microcomputer. Other conductors in an address bus are used to transfer 16 address bits together in parallel. The word length of a microprocessor is the

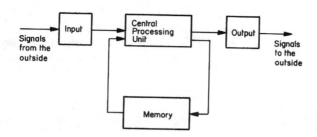

Fig. 9-1 Basic Components of a Microcomputer

number of data bits transferred together in parallel between the CPU and other components of the microcomputer. In our example, the number of data bits transferred in parallel together is eight, so the microprocessor has a word length of 8 bits. It is called an 8-bit microprocessor even though the address bits are transmitted in groups of 16. The greater the microprocessor's word length, the more data it can handle in a given time. The word length historically has increased from 4 bits to 8 bits to 16 and now 32 bits. This has resulted in an increase of power with each generation of microprocessors.

American Standard Code for Information Interchange (ASCII). English characters and Arabic numbers must be encoded in binary form before a computer can use them. ASCII is one of the most common of a number of codes that do this translation. ASCII is a seven-bit binary code with an eighth bit that is used as a parity check to detect errors. In one kind of parity the parity bit is a "1" if the seven bits add up to an odd number. If the seven bits add up to an even number, the eighth bit is a "0," as table 9-1 illustrates.

TABLE 9-1 SAMPLE ASCII CHARACTERS

English Character	ASCII Bit Position							
	1	2	3	4	5	6	7	8
G	1	1	1	0	0	0	1	0
L	0	0	1	1	0	0	1	1

In the letter G's case, the first seven bits add up to 4, an even number. Therefore, the eighth, or parity, bit is a 0. In the letter L's case, the seven bits add up to 3, an odd number, making the eighth, or parity, bit a 1.

General Organization of a Microcomputer

A microcomputer's organization, shown in figure 9-2, usually is referred to as the architecture. The heart of the system, the central processing unit (CPU) has two main functions: It controls the timing and sequence of operations in the computer, and it executes the arithmetic and logical operations of the data that are being processed. A memory unit stores the program or list of instructions known as read-only memory (ROM) to be executed. The memory unit also stores the data being processed, which is called the random access memory (RAM). To communicate with the outside world, input and output ports that allow data to be transferred to and from external devices such as printers, keyboards, display units, etc., are used. A system of bus lines ties together the microcomputer system components: the CPU, memory, and input/output ports.

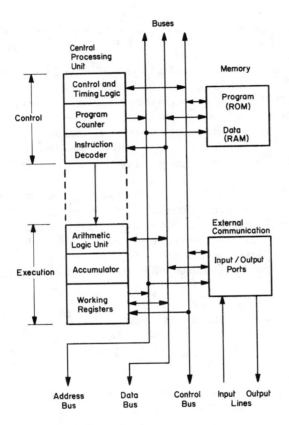

Fig. 9-2 Basic Architecture of a Microcomputer

Bus system. Data, control signals, and address transfers among the microprocessor's various units take place over sets of parallel wires known as buses. Most microprocessors have three buses: one for data, one for control signals, and one for addresses.

Control Bus. The control bus has a selection of signals to and from the CPU to control data timing and transfer on the other bus lines. One of these signals can halt the CPU and memory operation, disconnecting them from the bus system. This is used when more than one CPU in a microprocessor shares a bus, only one of which may access the bus at a time. A read-write line in the control bus determines the direction of the data bus.

Data Bus. The data bus transfers data to and from the CPU and the memory or input/output lines. This bus is bidirectional and controlled by the CPU. A read-write line in the control bus determines the direction of data in the data bus. If the CPU sets it to write, this signal allows the CPU to send data via the data bus to other parts of the system. When the control line is set to read, the CPU receives data from the data bus. Only one device may drive the data bus at a time, but all other devices may read data from the data bus at the same time. The data bus is tri-state in operation: 0 and 1 and a high-impedance state for disconnection.

Address Bus. The address bus is used to provide a signal for the memory to select one particular location or address within the memory to connect to the data bus. Smaller processors may have separate address lines for program memory and data memory. The address bus is like the data bus, tri-state in operation: 0 and 1, and high-impedance for disconnection.

CPU. It is important to understand the CPU's two main functions—timing and sequence-of-operation control and execution of operations—in detail.

Timing and Sequence-Of-Operation Control. This function is contained in the control and timing logic, the program counter register, a stack pointer, and an instruction decoder.

Control and timing logic—All of the microcomputer's different functional blocks must be timed or synchronized to work together. To assure accurate data transfer and proper operation, digital electronic circuits must have signals present, and signals must change at specific points in time. This is controlled by a clock signal. An example is the register, a device that can store 1 or 0 levels on its outputs to provide a temporary storage place for the digital codes as they are moved from one place to another in the system. A 4-bit register is shown in Figure 9-3. The input signal 1011 appears at the input in the form of a positive voltage for 1 and a negative voltage for 0. They are stored in the register until needed. The output does not change until the clock signal, which occurs at a regular time period, is applied. The clock signal controls not only registers but the entire system: control-line signals, address codes, input data, output action. These change only when a clock signal appears so the timing for all system parts is synchronized. Many of the microprocessors have an on-chip oscillator, which needs an external quartz crystal for the clock signal's timing. Also, there are different cycles the computer goes through, such as instruction fetch cycle and an execution cycle, that are controlled from the control and timing logic.

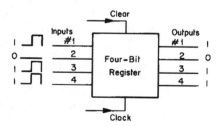

Fig. 9-3 Clock Signal and a Register

Program Counter.   The program counter's purpose is to give the memory address of the next instruction. The program counter is loaded with the memory address of the first instruction at the start of a program. When the system starts running, the address is transferred to the address bus and the instruction code is read into the instruction decoder from memory.

Instruction Decoder.   Part of the instruction is called operation code (opcode) and it specifies the operation to be performed. Typical operations are SUBTRACT, DIVIDE, ADD, and so on. Each has a different opcode. The instruction decoder then sets up all the logic linkages required to perform the desired operation. The actual performance of the operations is accomplished in the arithmetic-logic unit (ALU). A typical instruction execution consists of one or more instruction cycles. The first cycle, the instruction fetch cycle, calls in the opcode from memory to the instruction decoder. In simple instructions this is followed by an execution cycle when the required operation is carried out. If data are required, extra cycles are needed to read in the data words following the opcode. The timing varies with the type of microprocessor.

Subroutines—A subroutine is a small subprogram to accomplish a particular task. The main program activates a subroutine by a special instruction, such as CALL. The main program resumes on completion of the subroutine, which ends with a RETURN instruction.

Stack pointer—The stack is a special area of memory sites where information is stored in groups one after another to form subroutines. The last one to be written in the stack is the first one to be read out. The stack pointer is an address register that keeps track of the next instruction to be written in or read out. This order is sometimes referred to as last-in-first-out (LIFO). Other references call the stack order first-in-last-out (FILO), which means the same thing.

CPU Execution.   Execution involves the ALU, the accumulator, and working registers.

ALU—The ALU carries out the desired arithmetic or logic functions. The ALU has two data inputs and one output. The ALU provides such functions as ADD, SUBTRACT, AND, OR, EXCLUSIVE OR, COMPLEMENT, and CLEAR.

Accumulator—The accumulator is a special register that provides data for one of the ALU inputs. The ALU operation's results are placed in the accumulator after execution.

Working registers—Working registers are used to hold data or intermediate results.

Interrupt—An interrupt is an interruption of the main program by a peripheral, such as a printer or input keyboard. This interruption then switches the program to a subroutine that services the peripheral. The interrupt occurs when a peripheral requires data from the microprocessor or has data to give it. In a land mobile radio system an interrupt can briefly stop the main program while the peripheral is being serviced. This is especially true when there are a large number of peripherals.

Flags—Flags are individual bits that indicate the status of the last operation. An example is the C flag, which is set to 1 when an arithmetic operation produces a carry.

Address—An address is a binary pattern that represents a specific location in memory.

Addressable range—Addressable range is the number of memory locations the microprocessor can address.

Microcomputer memory.   Memory is basically divided into ROM and RAM.

ROM.   Figure 9-4 illustrates a ROM operation. The ROM stores a bit by the presence or absence of a connecting link between a row line and a column line in the memory array. The two address bits, A_0 and A_1, can use the decoder to select one of four rows to be read out. If both the A_0 and A_1 address lines have a logical 0, the decoder produces row 1 as an output. In this example row 1 is 1010, which appears at the ROM's output. Similarly, an address input of 0 for A_0 and 1 for A_1 produces an output corresponding to row 2, or 0100 in this particular ROM. A mask produces the open or closed link pattern in the ROM during the chip manufacturing.

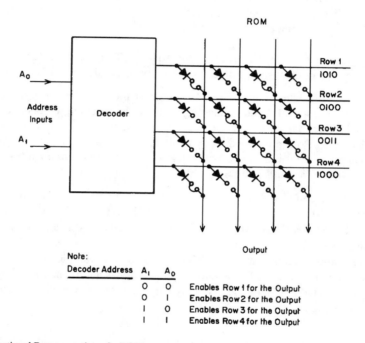

Fig. 9-4 Functional Representation of a ROM

RAM.   A semiconductor RAM is usually a volatile storage medium whose data are lost when power is removed from the chip. There are two basic types of random access memories: static and dynamic.

Static memory—A static memory stores the information in flip-flop or latch-type structures and maintains the information without requiring data refreshing or restoring.

Sec. 9.1 Reviewing Basic Applicable Computer Technology   **121**

Dynamic memory—A dynamic memory uses a storage medium such as a capacitor, which loses its information over a period of time and must be refreshed at intervals to ensure data retention.

Input/output. Digital data are transferred in and out of the microcomputer by way of data ports. The data ports consist of latched registers[*] connected to the data bus that the CPU can select for data transfers. Each input or output port has a pair of handshake, or control, lines. One handshake line goes from the computer to the external device, which indicates it is ready to transfer data via the port. The second line is an input from the external device to the computer system and may be used to indicate either that the external device is ready to accept data or that it has placed data on the port lines for the computer to read.

Modems. Modems are used to change the digital signal to audio tones for transmission on radio or telephone circuits. Modems also change the audio tones back to digital signals. The microcomputer's parallel digital signal output is usually changed to serial form before transmission through the modem.

Languages in the microcomputer. Three types of languages are important to land mobile radio systems: machine, assembly, and high-level.

Machine Language. The language of the digital circuits inside the computer is composed of 1 and 0 codes. The digital code the machine understands is called machine code. Instructions for the computer can be programmed directly into this code the machine understands. Programming directly in machine language is difficult and subject to many errors. To avoid this the computer is used to convert human commands to machine language. One of these methods is the assembly language program.

Mnemonic, or Assembly, Language. A mnemonic is an abbreviation of what the instruction does: LD for LOAD OPERATION, ST for STORE OPERATION, MOV for MOVE OPERATION, AD for ADD, SU for SUBTRACT, and so on. These mnemonics, or assembly language, are fed into an assembler in the computer, which converts the mnemonics into machine language. Table 9-2 shows an example of assembly language. C and D are called the operands. Each microprocessor has an instruction set in which the instructions are stated in mnemonic assembly language format, which the assembler converts into machine code.

TABLE 9-2 ASSEMBLY LANGUAGE SAMPLE

Instruction in English	Assembly Language	Machine Language
Move contents of Register C to Register D	MOV C, D	0100 0111

* A latched register is a type of temporary storage for digital information where the information is released on receipt of an electronic or digital signal.

High-Level Language. The mnemonic codes are not like normal human language. High-level language programs have been developed so programs can be written in a language closer to the human language. There are two methods of converting the high-level program to machine language. In the first a compiler reads the entire program and then translates it into machine language. In the second an interpreter reads each instruction one at a time and translates it into machine language. Some of the high-level languages land mobile radio systems use are BASIC, C, and Oracle.

Beginner's All-Purpose Symbolic Instruction Code (BASIC). At the present time the most common language for microcomputers is BASIC, invented by John Kemeny and Thomas Kurtz at Dartmouth College in 1964. It is used in a number of land mobile radio digital systems.

C Language. The programming language C was developed for use with the AT&T UNIX personal computer. It is used in CAD systems in land mobile radio.

Oracle. Some CAD programs involving interaction with data bases use Oracle.

Microprogram. A microprogram is a series of elementary instructions for computer operation stored in a ROM section in the microprocessor itself. Introducing a computer instruction into an instruction register initiates the microprogram's execution. The microprogram in its storage medium is referred to as a microcode.

Algorithm. A computer requires a precise sequence of instructions called an algorithm to perform a specific task.

Two synchronization systems. There are two general systems of synchronizing a serial system of bits transmitted and received: synchronous and asynchronous.

Synchronous. In synchronous the receiver clock continuously adjusts to the transmitter clock in frequency and phase. Data are sent at a fixed rate.

Asynchronous. In asynchronous the receiver clock does not adjust continuously to the transmitter clock. Data do not have to be sent at a constant rate. In computers start-stop bits for each character accomplish synchronization in an asynchronous format.

USART and UART. The universal synchronous/asynchronous receiver/transmitter (USART) is a general purpose synchronous/asynchronous interface chip that both types of communications use in a microcomputer system. If the interface chip implements asynchronous communication only, it is called a universal asynchronous receiver/transmitter (UART).

EIA/TIA-232-E. The EIA/TIA-232-E standard describes the interface between data terminal equipment and data circuit terminating equipment employing serial binary data interchange. Table 9-3 shows the pin connections for a 25-pin connector such as the Amphenol 17-10250-1, used to interconnect radio equipment with computers.

TABLE 9-3 EIA/TIA 232-E INTERFACE CONNECTOR*

Pin	Function
1	Shield
2	Transmitted Data
3	Received Data
4	Request to Send/Ready for Receiving
5	Clear to Send
6	Data Circuit Equipment Ready
7	Signal Common
8	Received Line Signal Detector
9	(Reserved for Testing)
10	(Reserved for Testing)
11	Unassigned
12	Secondary Received Line Signal Detector/Data Signal Rate Selector (Data Circuit Equipment Source)
13	Secondary Clear to Send
14	Secondary Transmitted Data
15	Transmitter Signal Element Timing (Data Circuit Equipment Source)
16	Secondary Received Data
17	Received Signal Element Timing (Data Circuit Equipment Source)
18	Local Loopback
19	Secondary Request to Send
20	Data Terminal Equipment Ready
21	Remote Loopback/Signal Quality
22	Ring Indicator
23	Data Signal Rate Selector (Data Terminal Equipment/Data Circuit Equipment)
24	Transmit Signal Element Timing (Data Terminal Equipment Source)
25	Test Mode

* By permission of the EIA.

Pipeline. Many of the 32-bit microprocessors use what is known as a pipeline: a series of stages from instruction fetching to instruction execution. Several instructions are in the pipeline at the same time. The first instruction may be at the instruction execution stage, while the second instruction may be at the operand fetching stage. Figure 9-5 shows the basic instruction pipeline concept. There are other types of pipelines, but the instruction pipeline is the most dominant.

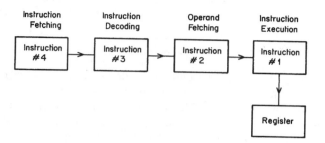

Fig. 9-5 Instruction Pipeline

Cache. A cache is a small, separate high-speed memory used for instructions or data that are needed often. This frees the main memory to supply other information. In several 32-bit microprocessors a cache is built into the microprocessor chip. The cache increases performance because the number of external fetches to external memory is reduced. An instruction cache increases performance by approximately 25%.

Operating Systems

A computer operating system provides the environment in which programs perform. A program properly developed for a standard operating system can be used on any computer using that operating system with a negligible amount of effort. These standard operating systems are discussed briefly:

Control program for microcomputers (CP/M). CP/M was the first standard operating system for microcomputers. It is a no-frill system. Its principal advantage is that different types of microcomputers can use it. CP/M is still in wide use.

Microsoft Corporation disk operating system (MS-DOS). There is more software available for MS-DOS microcomputers than for those that use any other operating systems. IBM calls the system PC-DOS (where PC stands for "personal computer").

UNIX. UNIX is AT&T's standard operating system, which is a much more flexible and powerful system than either CP/M or MS-DOS. It has advanced features such as allowing two or more users to share the same hardware simultaneously, and multitasking, where one user runs two or more programs at the same time. UNIX versions are available for a number of personal computers, such as the IBM PC and AT&T's UNIX personal computer.

OS/2. Developed by Microsoft and IBM, OS/2 is designed to replace MS-DOS. The new operating system has advanced features similar to the UNIX, including multitasking. The OS/2 is available only on computers that use microprocessors manufactured by Intel Corporation.

WINDOWS. WINDOWS is a new operating system Microsoft developed. The latest version is called WINDOWS NT where NT stands for new technology and is used for large, complex systems. It is a 32-bit system that can run several programs at once and has "crash" protection.

Database Management Systems

A database management system is a suite of programs that allows the user to store, retrieve, select, and associate information of various kinds in an efficient manner. It allows users to access the information in a computer's disks without knowing exactly where each piece of information is stored, and it makes it possible to see the relationships between different types of information without manual comparisons. dBase IV is an example of one database management system. There are other systems, and there is a standard syntax for retrieving information from databases kept in different types of computers. This syntax, developed by IBM, is called structured query language (SQL). Land mobile systems use geographical databases covering streets and similar material for CAD and automatic vehicle location.

Semiconductor Technologies in Microcomputers

There are basically four semiconductor technologies used in microcomputers: PMOS, NMOS, CMOS, and bipolar.

p metal-oxide semiconductor (PMOS). PMOS use p-channel field-effect transistors (FET) to implement logic functions. They operate from a negative voltage supply and sometimes require different voltage supplies. They were used in the earliest devices.

n metal-oxide semiconductor (NMOS). NMOS were used in later devices and require only a single positive voltage supply. NMOS are faster than PMOS but do require a voltage supply of ±5% tolerance. High-performance MOS (HMOS) is an improved version of NMOS in both increased speed and density, produced by scaling down device dimensions.

Complementary metal-oxide semiconductor (CMOS). CMOS use a combination of n- and p-type FETs. The CMOS have low power consumption compared to the other types. Radio systems use the CMOS microprocessors a great deal. CMOS is relatively unaffected by variations in the power supply and is less sensitive to noise on its input lines.

The original CMOS was a little slower in operation than the PMOS or NMOS. However, newer CMOS versions (RCA QMOS and Motorola HCMOS) are as fast as the NMOS. It should be noted that all CMOS types have high-impedance inputs and are susceptible to damage from static electric charges. Most of the devices have diode protection at the inputs to reduce the risk of damage. However, even with diode protection special care is recommended in handling these devices. This includes having the technician wear special grounding bracelets.

Bipolar integrated circuits. Bipolar integrated circuits are devices that operate by the transport of both electrons and holes. A bipolar transistor such as a conventional silicon transistor uses both electron and hole types of carriers. An FET is a unipolar transistor that uses charge carriers of only one polarity. Bipolar devices operate very fast but generally use up more space on the silicon chip.

Microprocessor packaging. Most of the microprocessors used in land mobile radio systems are packaged using the dual-in-line package (DIP). Figure 9-6 shows a DIP. There are a number of different types of packaging, but the DIP is the most popular. One of the new packaging techniques is surface mounting technology (SMT). This and other modern electronic equipment packaging are described in the References.

Fig. 9-6 Dual-In-Line Packaging

9.2 EXAMINING MICROPROCESSORS USED IN LAND MOBILE RADIO SYSTEMS

Motorola 6800

Motorola 6000 is an 8-bit microprocessor that came out in 1974. It has been used extensively in many land mobile radio systems. The 6800 DIP has 40 pins for external connections.

6800 buses. There are three buses: the data bus, the address bus, and the control bus. The data bus has eight conductors, which makes the 6800 an 8-bit microprocessor. The address bus has 16 address conductors, and the control bus has nine conductors.

6800 programmable registers. There are three 8-bit and three 16-bit programmable registers. Two of the 8-bit registers are accumulators that handle all the arithmetic operations. The third 8-bit register contains the flags, which are also known as condition codes. One 16-bit register is the program counter, which contains the address of the next instruction in line to be executed. The second 16-bit register is the stack pointer. This has the address of the top of the stack for writing in or reading out subroutines. The third 16-bit register is the index register. In an index register the register content is added to or subtracted from the operated address prior to executing an instruction.

6800 clock. A separate chip, MC 6875, is used to synchronize all the 6800 microprocessor's operations. An added external crystal governs the frequency of the MC 6875's internal oscillator.

Peripheral interface adapter (PIA). The PIA is a separate chip that connects the 6800 with peripherals such as printers and keyboards. The data to and from the peripherals are stored in the PIA registers until needed. The 6800 microprocessor uses these registers as a group of memory addresses. The 6800 has been supplanted by the 68000 and its family of microprocessors.

Motorola 68000

The 68000 is a 16-bit microprocessor with an internal 32-bit architecture. While it uses a 16-bit data bus, it has 32-bit user and internal registers and 32-bit operands. There are a number of packaging options for the 68000, but the DIP is the most popular. The 68000 DIP has 64 terminal pins for external connections.

68000 buses. The data bus has 16 conductors and the control bus has 25. The address bus has 32 conductors.

68000 programmable registers. There are 17 general-purpose registers, eight of which are data registers and seven that are address registers. Two of the address registers are used as stack pointers: the user stack pointer and the supervisor stack pointer. There is a program counter and a status register in addition to the general-purpose registers.

Status Register of the 68000. The status register holds variables indicating the status of the CPU and the interrupt system and the results of arithmetic.

User Stack Pointer. The user stack pointer is a register that has the address of the top of the stack for regular subroutines.

Supervisor Stack Pointer. The supervisor stack point register comes into play when interrupts and error conditions occur. The supervisor stack pointer is then used for special subroutines for these exceptions.

MC 68000 instruction set. There are approximately 59 mnemonics in the MC 68000 instruction set, each representing a separate operation code. Examples of the mnemonics and their corresponding operations are Bcc, Branch Conditionally; CMP, Compare; JMP, Jump; Move, Move. Each instruction in the MC 68000 consists of three parts: the operation code (mnemonic), a letter used to indicate the length of the operand, and the operand. The letter B (byte) is used for an 8-bit operand, W (word) is used for a 16-bit operand, and L (longword) is used for a 32-bit operand. An example of an instruction is

    MOVE. B <D2>, <D1>

This means: Move a copy of the first eight binary digits of source register D_2's content to destination register D_1. Since the internal registers of the MC 68000 can operate with eight, 16, or 32 bits, it is necessary to use B, W, or L to designate the operand's length.

Motorola 68008

The Motorola 68008 is an 8-bit version of the 68000. Sometimes in land mobile radio work only an 8-bit data bus is required for certain programs. The 68008 has an 8-bit data bus and it cuts the cost of a system. However, it still executes programs written for the 68000.

Motorola 68020

The Motorola 68020 is a full 32-bit microprocessor developed around 1983. It uses a 32-bit data bus and a 32-bit address bus. It has three 32-bit ALUs and a 32-bit-wide cache. The 68020 uses CMOS technology. The 68030 and 68040 are newer, faster versions.

Intel Microprocessors

The Intel 8080 is an 8-bit microprocessor used in land mobile radio systems. Although it functions as an 8-bit unit, it contains six 16-bit registers. The 8080 requires three separate power supplies and an external clock generator.

The Intel 8085 is an improved 8-bit microprocessor using one power supply and an internal clock generator. The 8085 time-multiplexes the data bits and the address bits on the same bus.

The 8085 was followed by the 16-bit 8086 and then the 32-bit 80386. The 486 is a newer 32-bit Intel microprocessor. The Intel 486 has 1.2 million transistors in a small chip. Intel now has a faster 32-bit microprocessor called Pentium.

Zilog Microprocessors

The Z80, an 8-bit microprocessor, is used extensively in land mobile radio digital systems. It has many similarities with the Intel 8080. However, the Z80 uses a single power supply and an internal clock signal. The Z80 can use the available 8080 software. The 16-bit Z8000 succeeded the Z80. A 32-bit microprocessor, the Z80000, followed the Z8000.

9.3 COMPARING ADVANTAGES AND DISADVANTAGES OF MICROPROCESSOR CONSOLE CONTROL

In a large dispatching center a number of consoles are used for both radio control and taking telephone calls. The decision is whether to use microprocessor-controlled consoles. Each case must be judged on its own.

An evaluation of advantages and disadvantages requires a knowledge of switching architectures. There are two main systems: analog and digital.

Analog voice switching using a centralized switch. Essentially, analog voice switching uses one system controller and a backup for all the consoles. The system controller contains a microprocessor that controls all the switching. In addition, there is a microprocessor for testing. This makes a total of three microprocessors for the entire console system.

Digitized voice switching. The voice at each console is digitized and then time-multiplexed on one bus that connects all consoles. A console has a number of microprocessors, one for each transmitter. There is no single microprocessor controlling the entire system.

Advantages

1. Microprocessors give a high degree of operator convenience using cathode ray tube (CRT) and keyboard. This includes using colors to enhance operations and displaying all information in one compact area. The microprocessor makes an integrated display of automatic number identification (ANI), messages, directories, and so on possible.
2. Using more streamlined procedures improves operator efficiency. A microprocessor can preselect an all-points bulletin in police work, have a built-in speed dial for making phone calls, and automatically set up an incoming call queue.
3. Console features can be added, which would be difficult to implement without a microprocessor. An example is a single-tone decoder's selection of a free operator. In addition a microprocessor enables the supervisor console to push the proper button and take over the function of any other console.
4. Specials to be installed at a later date may be accomplished by software. This results in less wiring and less special hardware when making modifications. Microprocessors make the console more flexible in future operations.

Disadvantages

The disadvantages relate primarily to reliability.

1. Centralized decision-making could result in catastrophic system failure. Using a number of microprocessors can minimize this possibility.
2. Software has reliability problems.

9.4 COMPUTER-AIDED DISPATCH SYSTEMS

CAD systems are used in large dispatching organizations such as fire, police, and EMS to assist the dispatcher. For example, in a large fire department the CAD system recommends which fire apparatus and how many pieces should be dispatched to a particular fire. This is done with the aid of computers that have stored lists of street addresses and alarm box numbers in their memories. Algorithms determine the nearest firehouses to any address and determine the number of units dispatched from information sent from the fire scene. In addition, the computer system can recommend a list of fire units available for relocation to cover the firehouses emptied by dispatching to a fire. CADs do, of course, differ for each installation. However, to demonstrate what such a system can do, here is a simplified description of a specific system that has been in operation for a number of years.

CAD System in the New York City Fire Department

Each of the five boroughs in New York City has its own fire central office (dispatching center) and its own radio system.

Computer setup. The computer setup shown in figure 9-7 includes the borough computer and a central computer. The borough computer is a concentrator, holding all the inputs and then sending the borough information to the central computer, which does all the dispatching recommendations and sends them back to each borough.

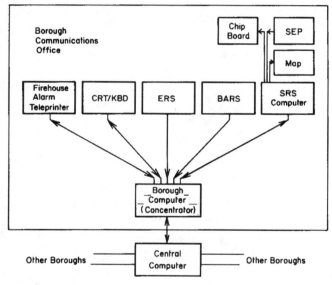

ABBREVIATIONS:

SEP – Status Entry Panel (Type in Status Information)
SRS – Status Reporting System Computer
BARS – Box Alarm Readout System (for Mechanical Alarm Boxes)
ERS – Emergency Reporting System (for Voice Alarm Boxes)
CRT – Cathode Ray Tube / Keyboard (Input and Display Alarm Information)

Fig. 9-7 Computer Arrangement in a Fire Department CAD System

Status of fire vehicle apparatus. In each borough's central office there is a map showing the status of fire vehicle apparatus in that borough. In a citywide fire command center there is a map showing the status of all fire vehicle apparatus in the city.

There are three colors: green—unit available on air or in firehouse; white—en route or at nonserious incident; red—out of service, operating at serious incident, or relocated.

Backup status board (chip board).   The chip board is a backup mechanical system that goes into operation when the computer is down. The colors for this are green—unit in firehouse; amber—unit available on the air; and red—unavailable.

Radio operator.   As the fire alarm flow diagram in figure 9-8 shows, the radio operator dispatches units that are not in firehouses but are available on the radio. The radio operator does this by observing the highlighted units on the screen. These are automatically sent to the radio operator when the decision dispatcher releases the computer-recommended-dispatch.

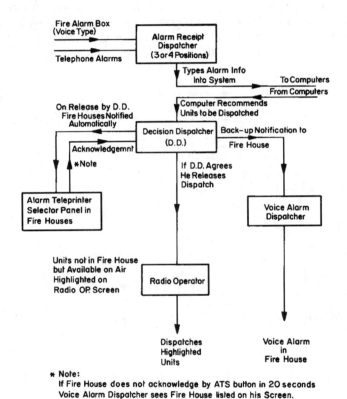

* Note:
If Fire House does not acknowledge by ATS button in 20 seconds
Voice Alarm Dispatcher sees Fire House listed on his Screen.

Fig. 9-8 Fire Alarm Flow in a Borough Central Office

When the first unit reaches the fire, it radios back a 10–84 signal to the radio operator. The radio operator then types the box number (number of alarm box), the unit, and 10–84 on the status entry panel (SEP). Similarly, the radio operator types in other information received on the radio such as 10–92 (false alarm). When the radio is busy, there is a radio-in dispatcher who types in the information on the SEP and radio-out person who does the radio transmissions.

About a minute after the original dispatch, the radio operator makes an additional announcement of the fire, including type and address. After the original radio dispatching, the radio operator pushes ACK. (acknowledge), then Clear or Next on SEP. "Next" gives the radio operator the next incident or blank screen if there is no other incident. The computer prints out a ticket of each incident, which the radio operator uses to write on when the radio receives the information.

Available screens for the radio position. Each position—alarm receipt dispatcher, decision dispatcher, supervisor, radio, and so on—has a number of available screens that can be called up on the CRT. All positions except the alarm receipt dispatcher (ARD) have two CRTs. One is a summary screen (summary of incidents currently active) and the other is a working CRT with a number of screens. The available screens for the radio operator on the working CRT are

1. *Alarm inquiry.* Page 1 displays the alarm assignment card for a user-specified alarm box for the first alarm through the fifth alarm. Page 2 displays the location of the user-specified box on a borough outline map.
2. *Dispatch notification—radio.* The dispatch notification screen lists units (highlights) dispatched to an incident that are to be notified by the radio operator.
3. *Idle screen.* The idle screen is used for entering line-24 command, calling up other screens.
4. *Incident history.* The incident history screen gives a complete minute-by-minute history from the first alarm, each unit assigned, to the end of the incident. It can be invoked by entering line-24 command plus an incident number on idle screen.
5. *Message receipt screen.* The message receipt screen is used to receive messages. It is invoked by pressing the Next button on the CRT in response to an "MSG wait" light.

Digital status reporting system (DSRS). A second-generation CAD system in the New York City Fire Department uses Mobile Data Terminals in vehicles. These transmit data about vehicle status directly on 800 MHz to a computer, eliminating the need for the radio operator in the borough central office to type in vehicle status on the SEP. Sometimes DSRS is called mobile status entry systems (MSES). Figure 9-9 shows the integration of a second-generation CAD with a mobile data system. This is a true combination of computer technology and land mobile radio systems.

Modems at the base station change the incoming data radio signals into information bits that go to an interface computer and then are translated into the format the CAD computer uses. Similarly, modems at the base station change the outgoing computer signals into a form radio can transmit to the mobile data terminals and computers.

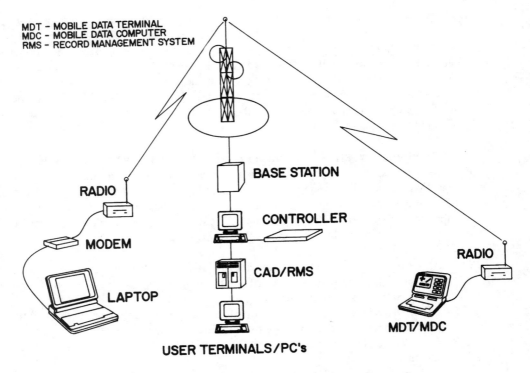

MDT - MOBILE DATA TERMINAL
MDC - MOBILE DATA COMPUTER
RMS - RECORD MANAGEMENT SYSTEM

BASE STATION

RADIO

CONTROLLER

MODEM

CAD/RMS

RADIO

LAPTOP

MDT/MDC

USER TERMINALS/PC's

Fig. 9-9 CAD Integrated into a Mobile Data System
Permission of The Warner Group, Management Consultants

REFERENCES

Brenner, David J., and Marilyn A. Cadoff. "Guide to Computer-Aided Dispatch Systems," *NBSIR 84–2991*. Gaithersburg, MD: U.S. Department of Commerce, National Bureau of Standards, Law Enforcement Standards Laboratory, 1984.

Ginsberg, Gerald L. *Electronic Equipment Packaging Technology.* New York: Van Nostrand Reinhold, 1992.

Greenfield, Joseph D. *Microprocessor Handbook.* New York: John Wiley & Sons, Inc., 1985.

Harmon, Thomas L., and Barbara Lawson. *The Motorola MC 68000 Microprocessor Family.* Englewood Cliffs, NJ: Prentice-Hall, Inc., 1985.

10

Taking Part in the Digital Revolution

The telephone network is being redesigned into an integrated services digital network (ISDN) that will provide a complete digital connection throughout the country. The ISDN will support a wide range of services, including computer interconnection, data transmission of all kinds, and other nonvoice and voice services. This transmission of data in the telephone network is being paralleled in the land mobile radio services: mobile digital radio dispatch systems, mobile graphics, digital status reporting systems, computer-to-computer communications via land mobile radio, and so on, for example.

This chapter explains the difference between bits per seconds and bauds. It examines the factors that determine how many bits per second can be sent over a given radio channel. Examples of synchronizing digital radio receivers with transmitters are given. Methods to detect errors in two-way digital radio systems are discussed, and the different types of digital modulation are examined. The components of a digital radio communication system are described, including the functions of a network processor. The different methods of channel control in digital radio systems are evaluated.

Using packet technology with a conventional two-way radio system is described, as are digitized voice cellular systems.

Finally, APCO Project 25 is developing a new technology standard for digital radio including data and digitized voice. The common air interface of this standard is described.

The change from analog to digitized voice is a major one. Three strategies for dealing with this complete rapid displacement of equipment are discussed: backwards-compatibility, dual mode operation, and software determination of equipment characteristics.

10.1 COMPARING BITS AND BAUDS

The most important feature of digital communication is that only a defined number of symbols are transmitted. If the information is binary, only two symbols are needed, a 1 or a 0. These are called bits (binary digits). Each bit defines one of two conditions: yes or no, on or off, etc. The number of symbols per second is called a baud. In this example the number of bauds is equal to the number of bits per second.

In a set of four symbols each symbol defines one of four states. An example is 00, 01, 10, and 11 where each state is a group of two bits. In this case the number of bauds is one-half the number of bits per second.

10.2 BINARY VERSUS MULTISYMBOL SYSTEMS

The use of multisymbol codes is called m-ary signalling. A binary signal is more easily and accurately detected in the presence of noise. On the other hand increasing the number of possible symbols increases each symbol's information content. And multisymbol systems allow transmission of more information per second for a given channel bandwidth. Many modern digital radio systems use four-state or symbols to increase the information rate.

10.3 EXAMINING A DIGITAL RADIO CHANNEL'S CAPACITY

In 1949 Shannon showed that the maximum possible error-free bit rate C over a channel of bandwidth W is given by:

$$C = W \log_2 (1 + S/N)$$

where

C is b/s
W is bandwidth in Hertz
S is mean signal power
N is mean noise power

For example, in a binary system where W is 6 kHz and the signal to noise power ratio is 15 (equivalent to 11.7 dB signal-to-noise), the maximum theoretical channel capacity is 24,000 b/s. In practice there are a number of factors such as guard-band requirements, multipath fading, and interference that reduce the actual channel capacity.

10.4 SYNCHRONIZING DIGITAL RADIO RECEIVERS WITH TRANSMITTERS

Digital radio receivers must be synchronized with their transmitters to receive digital communications. There are two general synchronizing methods: asynchronous and synchronous. One asynchronous method used in digital radio paging employs a series of 1s and 0s at the beginning of a short transmission for bit synchronization. This is followed by a sync word to obtain word synchronization. The system is described in detail in Chapter 11.

In synchronous data transmission a transmitter clock sends out data at a uniform rate to a receiver that clocks in the data at an identical rate. The data are transmitted in blocks of characters. Each block is of the same length and has special sync characters that identify the start and end of the block of characters.

In one synchronous radio digital system the transmitted blocks of characters are called frames. At the beginning and end of a frame there is a sync character called a flag. Each flag is one byte or 8 bits and denotes the beginning or end of a frame. This is the system used in the packet AX.25 protocol, which has been adapted from the data industry.

10.5 DETECTING ERRORS

Digitally encoded information is transmitted by sending a stream of pulses or bits. Fading and impulse noise can introduce errors, which cause problems unless they are detected.

Intersymbol interference occurs when pulses are spread in time to affect the signal in adjacent time slots. Multipath propagation can cause this.

In two-way digital radio systems the receiving station detects the errors. The receiving station then does not send an acknowledgment and the message is repeated. There are two basic ways errors are detected; parity bits and cyclic redundancy checks.

Parity bits. A parity bit is added to the end of a binary word. In even parity a parity bit is added so that the total number of bits is even. (In odd parity a parity bit is added so that the total number of bits is odd.) An example of even parity: a 0 bit is added to the original word 1 1 0 1 0 0 1 to form 1 1 0 1 0 0 1 0. The number of ones add up to four, an even number. In odd parity a 1 would be added to the original 1 1 0 1 0 0 1, making it 1 1 0 1 0 0 1 1. The number of ones now add up to five. This simple parity scheme can detect one bit error but not two. In addition this scheme uses up bits for error detection. The CRC is a more efficient error detection scheme used in two-way digital radio systems.

Cyclic redundancy check. In CRC the bits of the frame are transmitted with a special two-character sequence added at the end. This two-character sequence is obtained at the transmitter by sending the data through a special 16-bit shift register. At the receiver a similar 16-bit shift register treats the received data in the same way, also producing a two-character sequence. The two sets of two-character sequences are compared at the receiver. If they are the same, an acknowledgment is sent. If they are not, the transmitter repeats the message. There are variations in different CRC systems, but the principle is the same: The CRC method checks each frame for errors. CRC is sometimes called check sum.

10.6 FORWARD ERROR CORRECTION (FEC)

FEC is the correction of errors at the receiver. Most radio digital circuits accomplish FEC by using special codes to correct errors the multipath transmission of signals and burst noise cause.

A number of these codes are used in digital radio land mobile communications. For a more comprehensive treatment of this complex subject see the References at the end of this chapter.

Hamming codes. Table 10-1 shows a simple hamming code.

TABLE 10-1 HAMMING CODES

Information Bits	Added Bits
1010	101
1000	110
1100	011
1001	001
1011	010

In hamming every code word has seven digits, of which four are information bits. Each seven-digit code differs from any other by at least three positions. This allows a single-bit error to be corrected. A microprocessor compares each received code word with a stored list of correct code words. The microprocessor searches the stored list for a code word that gives the minimum bit difference. If there is a one-bit error in the received code word, the microprocessor selects the unique stored code that has a one-bit difference. In this example the hamming difference is three because each code word differs from any other by at least three positions.

BCH codes. Bose-Chaudhuri-Hocquenghem (BCH) codes are a generalization of hamming codes that allow multiple error corrections. They provide a large class of easily constructed codes of arbitrary block lengths. The BCH_{nkd} code has n total bits in a code word, of which k are information bits and d is the hamming distance.

Golay codes. Golay codes are a special type of BCH codes that allow triple error corrections. Golay codes, like BCH codes, are binary.

Reed-Solomon (RS) codes. RS codes are an important subclass of BCH codes. They are defined in terms of a Galois field: a finite field designated GF_q where q is a finite set of elements. In general a finite field exists only when the number of elements is a prime number or the power of a prime number. Finite fields such as the Galois field have their

Chap. 10   Taking Part in the Digital Revolution

own rules of arithmetic. For each value of q there is only one unique field. In RS codes the block length is $n = q - 1$. These codes correct multiple errors. RS codes use groups of six bits called hex bits.

Convolution, or trellis, codes. A trellis is a diagram used to represent the state of a convolution encoder. A number of shift registers and two or more adders wired in a feedback network generate convolution codes They are called convolution codes because the encoder's output is the convolution of the incoming information bit stream and the bit sequence that represents the impulse response of the shift register and its feedback network. As each incoming information bit propagates through the shift register, it influences several outgoing bits, spreading the information content of each data bit among several adjacent bits. An error in any one output bit can be overcome at the receiver without losing any information.

Convolution codes are classified by their code rates. The code rate is the ratio of the number of input symbols going into the convolution encoder to the number of output symbols. One-half and 3/4 are typical rates.

10.7 DIGITAL MODULATION SYSTEMS

There are three basic methods of modulating the RF carrier in digital land mobile radio systems: frequency shift keying, phase shift keying, and amplitude shift keying. There is also a combination of phase shift keying and amplitude shift keying.

Frequency shift keying (FSK). FSK is a change or shift in the RF carrier frequency to represent 1 in the binary code. A different amount of change in the RF carrier frequency is used to represent 0 in the binary code. Frequency modulating an RF oscillator or switching between two RF oscillators can generate FSK. The receiver has two filters, each tuned to one of the FSK frequencies. FSK is the most popular method of RF carrier digital modulation.

There are a number of FSK variations used in narrowband transmission. Minimum shift keying (MSK) is when the modulation index is fixed at 0.5 and the frequency shift is kept to 0.25 times the bit rate. GMSK is MSK with an added premodulation filter such as a Gaussian low pass. Another technique is tamed frequency modulation (TFM), which uses special premodulation filters to control spectral spreading. There are also four-level FSK modulation systems. For example, the binary number 00 represents a carrier frequency deviation of +0.50 kHz, 01 represents a carrier frequency shift of +1.50 kHz, 10 represents a carrier frequency deviation of –0.50 kHz, and 11 represents a carrier frequency deviation of +1.50 kHz.

Phase shift keying (PSK). PSK in general is more complex and expensive than FSK. However, PSK has good noise rejection and is used at higher bit rates such as 9600 b/s.

Biphase shift keying (BSK) is a two-phase level form of PSK. This uses a phase shift of the carrier for 1 in the binary code and another phase shift 180° apart for 0. The biphase form is designated 2-PSK.

Quadrature phase-shift keying (QPSK) is a four-phase form of PSK, designated 4-PSK. For example, the binary number 00 represents a phase shift of 90°; 01, a phase shift of 180°; 10, a phase shift of 270°; and 11, a phase shift of 360°. $\pi/4$ QPSK is a form of QPSK where the four phase shifts are +45°, −45°, +135°, and −135°. Differential phase-shift keying (DQPSK) is a form of phase shift keying in which the reference phase for a given keying interval is the phase of the signal during the preceding keying interval. $\pi/4$ DQPSK is a variation of DQPSK used in high-speed digital systems. Another variation of $\pi/4$ QPSK is $\pi/4$ QPSK-VP where a time varying phase-waveform is imposed on the $\pi/4$ QPSK time slot. This improves the bit error rate (BER) performance, and it is used in some in-room transmissions. Four-phase shift modulation is attractive because of its low error rate in a noisy environment.

Amplitude shift keying (ASK).   In ASK the amplitude of the modulating signal determines whether a 1 or a 0 is being transmitted. This in itself is not used much in land mobile radio systems. However, it may be used in combination with phase shift modulation.

Quadrature amplitude modulation (QAM).   QAM is a hybrid of 4-ASK and 4-PSK systems. QAM's advantage is that it can transmit more information in a radio channel than QPSK.

Direct digital modulation.   Direct digital modulation is when the previous methods are used to modulate an RF carrier.

Indirect, or subcarrier, modulation.   Indirect modulation is the use of digital modulation of the baseband. The data signal then is treated as a voice signal using the transmitter's analog AM or FM modulators. This system is more compatible with existing analog transmission facilities. It is often preferred for low speed rates. Direct modulation uses less than half the bandwidth of indirect modulation and thus is preferred with high bit rates that have to be transmitted over a given channel.

10.8 USING DIGITAL COMMUNICATIONS IN LAND MOBILE RADIO SYSTEMS

The radio digital communication system described here differs in a number of general ways from the conventional voice radio system. First, everything is controlled by computers, usually microprocessors. Second, the receiver voting system, if present, does not depend on a comparator to select the "best" signal. Instead the system selects the first correct signal from the voting or satellite receivers. Another difference is that in digital systems some form of synchronization between the transmitter and receiver is necessary. In addition error detection and correction are used in digital systems to ensure accurate communication. Figure 10-1 shows a generalized block diagram of a two-way radio digital communication system. There are many variations, depending on the particular use.

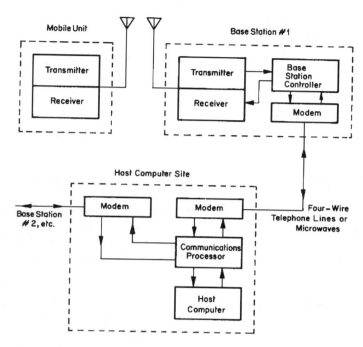

Fig. 10-1 Basic Land Mobile Radio Digital System

Parts of a Land Mobile Radio Digital System

The following are the basic components of every land mobile radio digital system:

Mobile unit. The mobile unit in general consists of a transmitter and a receiver connected to an antenna. In addition there is an MDT that generally includes a keyboard, a display unit, and a microprocessor. The display unit shows alphanumerics and in some cases graphics. Figure 10-2 shows one configuration. Each mobile unit has an identification number that is automatically sent with each message.

The MDT can be operated as a computer terminal. It can query a host computer for information or it can input information into a host computer. The keyboard may have user-programmable keys and a standard (QWERTY) typewriter keyboard for free-form messages.

Base stations. One base station or a number of base stations can operate in a network. Each base station has a transmitter and receiver, a base station controller, and a

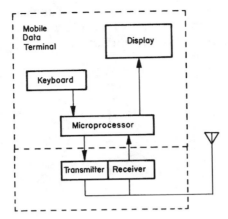

Fig. 10-2 Mobile Unit for Digital Communications

modem to connect to a telephone line (or microwaves) for transmission to a host computer. The base-station controller is connected to the base station. The base-station controller recovers the data, decodes the message, performs error detection and correction, and prepares the data for transmission to the communications or network processor. In some controllers there are two microprocessors. One supports the RF communications functions and the other supports the functions connected with the communications or network processor.

Satellite receivers. The receiver in each base station acts as a satellite receiver, picking up mobile signals. The communications or network processor selects the first correct message. In some cases satellite receivers alone are installed at sites to increase mobile coverage. In these cases each satellite receiver is connected to a controller that functions like the base-station controller. The communications or network processor again selects the first correct message received.

Communications or network processor. When a number of base stations make up a system, the processor that controls the system usually is referred to as a network processor. With one base station in a system, the processor usually is called a communications processor. In general the functions are:

1. Acknowledge the successful reception of a message to the mobile unit and forward the message to the host computer through an interface.
2. If the same message is received from a number of receiver sites, the processor selects only the first one.
3. Report to the host computer system with the status of messages that are undeliverable to the mobile data equipment.
4. Report to the mobile data equipment the status of messages that cannot be delivered to the host computer.

5. Control the RF channel protocol, which is the means to control the orderly communication of information between the base station and the mobile units.
6. If the system has more than one base station, the processor prevents two base transmitters from operating at the same time.

Half-duplex (HDX) and full-duplex (FDX) in data systems. HDX devices can send and receive data but not simultaneously. FDX devices can send and receive simultaneously. Terminals and modems can be defined by whether they are HDX or FDX.

Automatic repeat requests (ARQ). The base transmitter usually acknowledges the mobile transmission. If the mobile does not receive the acknowledgement, the mobile automatically repeats the message a number of times.

Channel throughput. Channel throughput is the number of correct messages per hour the system can process.

Vehicles per channel. To calculate the vehicles per channel, it is necessary to know the total throughput in correct messages per hour that the system can process, T. It is also necessary to know the average number of messages per vehicle per hour, A. Then the vehicles per channel can be calculated by $N = T/A$.

Channel control. Most radio digital systems use data burst transmission known as packets. Several methods are used to prevent stations from transmitting at the same time and jamming each other.

1. *Contention control.* All units transmit at random. This is called pure Aloha, as it was named in a report by the University of Hawaii. Since transmissions can collide, contention control yields a maximum throughput of 18% of channel capacity.
2. *Polling scheme.* In a polling scheme each terminal is asked in sequence whether it has anything to transmit. This system is not used much because of the time it takes.
3. *Carrier sense multiple access (CSMA).* An external RF carrier's presence is used to inhibit the transmission of a message. This prevents transmission when the channel is busy. Theoretically, the channel efficiency is 100%. However, in practice there are limitations such as station busy detect time (about 100 ms) and transmitter attack time.
4. *Digital sense system.* In a digital sense system the base station continuously transmits a bit pattern that is changed when the channel is busy. The digital busy indicator is set when an inbound signal is detected at the base-station receiver. The signaling delay is limited primarily by the transmitter attack time.

Types of Land Mobile Radio Systems

Computer to computer via two-way land mobile radio. Motorola has designed and built a computer together with a two-way radio in the same portable unit. IBM service personnel originally used it to communicate with a host computer through a network similar to figure 10-1. The RF transmission is at 4800 b/s with automatic channel sensing. The land line transmission is at 2400 b/s. The RF channel uses an error correction code that accounts for the difference in signaling speeds between the RF and land line channels.

Over 1000 portable computer terminals can operate on a single radio channel. The portable computer unit has an alphanumeric keyboard and a 54-character liquid-crystal display. It holds 80 kilobytes of RAM and 160 kilobytes of ROM. The portable unit uses two 8-bit microprocessors. The network control processor uses a Motorola 68000 microprocessor. Today this and similar systems are in general use for computer-to-computer communications via two-way land mobile radio.

Digital radio dispatching systems. Mobile Data International has developed a digital radio dispatching system based on the Zilog Z80 microprocessor. This, too, is similar to figure 10-1. The mobile units use modified standard mobile radios. Other companies, such as Motorola, design several types of digital radio dispatching systems:

Digital Dispatching with Graphics Display. The system the fire department of Phoenix, Arizona uses is one of digital dispatching with graphics display. Graphic information such as maps of routes, drawings of buildings, locations of hydrants, locations of stairways, and so on, are transmitted from a host computer to an MDT, where they are displayed on a 5-in. CRT. This is part of a computer-aided dispatch system.

Computer Query. A number of police jurisdictions have MDTs that can query a base computer for information on auto licenses. Such information may concern ownership (stolen car), traffic or parking violations, and so on.

Digital Status Reporting System (DSRS). DSRS also is referred to as mobile status entry system (MSES). Pushing a status function key in the mobile unit transmits the status of the vehicle in digital form to a base station and then to a host computer that is part of a CAD system.

Mobile Digitized Voice Terminals. Experimental work is being done on converting speech into digital signals to use with mobile digitized voice terminals. A continuously variable voice delta modulation system is used with linear predictive code vocoders (voice coder). A vocoder is a band compression speech system. Delta modulation is a method of converting analog information into digital pulses. The voice signal is quantized and binary pulses are produced that carry the information corresponding to the derivative of the modulating signal's amplitude. That is, instead of the absolute signal amplitude being transmitted at each sampling, only the changes in signal amplitude from sampling instant to sampling instant are transmitted.

Mobile Typed Digitized Terminals. A message that is typed, converted to a digital form, and transmitted by radio can be converted to voice by a voice synthesizer at the receiving end. Speech can be synthesized by combining phonemes, which are the building blocks of speech. A single chip converts a digital code into phonemes. Approximately 64 different phonemes are used.

10.9 USING PACKET RADIOS INSTEAD OF TELEPHONE LINES FOR DATA TRANSMISSION

Most radio digital communications are transmitted in bursts called packets.

Packet Protocol

The packet protocol the data communication industry uses is called the CCITTX.25. A modification of this protocol radio amateurs and others use is the AX.25. Figure 10-3 shows a functional representation of one frame of an AX.25 packet.

There may be a number of frames separated by flags. Each flag is a sync character one byte in length that denotes the beginning or end of a frame. The address field contains the recipient address and the addresses of any required digipeaters (repeaters). The control field controls the flow of packets and messages in their correct order. The data field contains a message of up to 256 ASCII characters. The frame check sequence (FCS) or CRC, is a method of error checking.

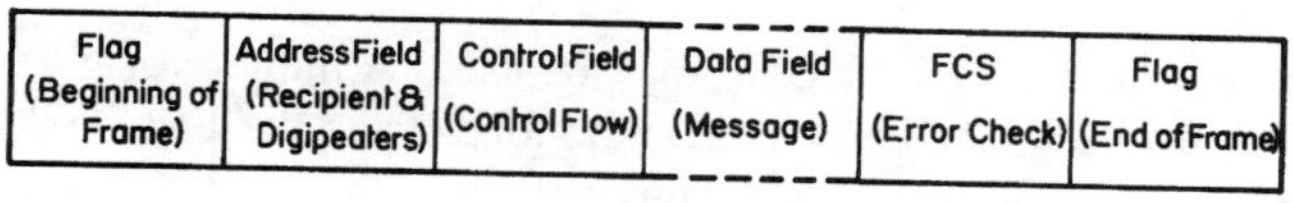

Flag (Beginning of Frame)	Address Field (Recipient & Digipeaters)	Control Field (Control Flow)	Data Field (Message)	FCS (Error Check)	Flag (End of Frame)

Fig. 10-3 Functional Representation of One Frame of AX.25 Packet

Packet radio systems. Packet radio systems now are used in place of telephone lines to transmit digital data, thanks to telephone lines' rising cost and line noise. This usually is done by adding a radio packet controller or transmitter node controller (TNC) to an existing two-way radio system. Figure 10-4 shows a packet radio system that consists of a computer with a printer, a TNC, and a two-way voice radio.

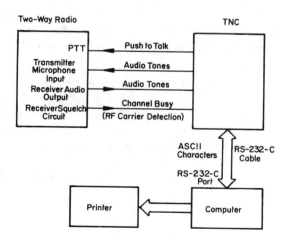

Fig. 10-4 Packet Radio System

Computer.   The computer may be a PC, but it must have an RS-232-C serial asynchronous port. The computer also must have communications software for encoding and decoding ASCII characters and for sending and receiving files that contain the data. This communications software is the same as for a telephone modem. The computer is connected by a 25-wire RS-232 cable from the computer RS-232 port to the TNC. Actually, a nine-wire cable is sufficient using pins 1, 2, 3, 4, 5, 6, 7, 8, and 20. (The RS-232 pin functions are shown in table 9-1.)

TNC.   The TNC consists of a microprocessor and memory, a high-level data link control (HDLC) chip, and a modem. The TNC strips the ASCII characters of start-stop and parity bits. The HDLC chip assembles the packet shown functionally in figure 10-3. Frame flags delineate the beginning and end of each frame of a packet. There are no start-stop bits at the beginning and end of each character and the packet is synchronous.

The packet is then sent to the modem. The modem changes the bits into two audio-frequency shift keying (AFSK) tones—one for a mark and one for a space. The tone format may be Bell 202, with 1200 Hz for a mark and 2200 Hz for a space. Other modem tone formats used are CCITTV.21 and CCITTV.23. The baud rate for radio transmission is usually 1200 for VHF and UHF.

Radio.   A two-way voice radio on VHF or UHF may be used. A number of wires connect the radio to the TNC. One wire sends the modem audio tones from the TNC to the radio transmitter's microphone input. The microphone input still can be used for voice transmission. A second wire connects the TNC to the radio transmitter's PTT connection. A third wire connects the receiver audio to the modem in the TNC. Here the audio tones are converted back to mark and space. A fourth wire is used to inhibit transmission when an external RF carrier is present.

Recovering the data message.   The TNC strips the data message from the packet and adds start-stop and parity bits to the ASCII characters. The asynchronous ASCII characters are then sent to the computer, where the message is read out and printed.

Transmission Speed.   At 1200 bauds approximately 120 characters per second are sent. The actual message transmission speed is less because of the other packet information that is transmitted with the message data.

Channel Access Control.   There are two techniques packet radios use to inhibit packet transmission when another packet is being transmitted on the same channel. The first method operates by detecting another packet's audio. This does not work when voice is on the channel. In that case RF carrier detection or CMSA must be used. (This is the method illustrated in figure 10-4.) In a special exception one station's packet is initiated at exactly the same time as another station's on the same channel. The two packets collide and neither gets through. Each station then retransmits at a randomly determined time to avoid a second collision.

Digipeaters. Digipeaters relay digital packets. They may operate on one frequency, storing the received signal and then retransmitting the message. As many as eight digipeaters have been used to relay packets a considerable distance.

10.10 DIGITAL CELLULAR SYSTEMS

Digital cellular converts human voice into a digital format that is transmitted through the cellular network. Analog cellular was discussed in Chapter 7.

Changing voice from analog to digital. Digital voice requires a frequency range of three to four times the analog voice's channel bandwidth for the same voice quality. Reducing the channel bandwidth by decreasing the bit rate decreases speech and speaker recognition. The quality standard in many cable transmissions is 64 kb/s for a digitized voice channel. This bit rate is too high for digital cellular because it requires a large channel bandwidth. Special digital compression methods are used in different systems.

Digital cellular advantages.

1. *Increased capacity.* Digital mobile receivers are more tolerant to interference, which means that base transmitters can transmit with less power. Therefore, all co-channel cells require less separation, which allows a much greater capacity than the analog system's.
2. *Lower power consumption.* Digital portable units require smaller batteries. This makes smaller portable units practical.
3. *Lower cost.* Digital systems are generally lower in cost.
4. *Compatible to ISDN.* The compatibility extends many new services to the mobile cellular phone system.

TDMA digital cellular. TDMA digital cellular is one of the systems now in use in the United States. Ericsson G.E. employs TDMA digital cellular systems supporting initially three users per carrier at a bit rate of 48.6 kb/s. It plans to go to six users per carrier. EIA/TIA/IS-54-B, a cellular system dual-mode mobile station-base station compatibility standard, forms a compatibility standard for cellular mobile telecommunications systems that are incorporating TDMA digital technology in the United States. The standard accommodates both analog and TDMA mobile units in the same system. Japan is introducing a TDMA system conceptually similar to IS-54. In Europe there is a Pan-European TDMA cellular mobile radio system known as GSM, named after the Group Schedule Mobile Committee that guided its development and was responsible for its specification. The GSM network supports eight TDMA channels per carrier at 270.8 kb/s. The GSM standards information can be obtained from European Telecommunications Standard Institute (ETSI). ETSI Publications Office, Boite Postal 152, Sophia Antipolis, 06561 Valbonne Cedex, France. Fax: +33 93 65 47 16.

CDMA digital cellular. CDMA digital cellular is another system now in use in the United States. Qualcomm manufactures this type of digital cellular based on spread-spectrum technology. An intermediate EIA/TIA Standard IS 95 for CDMA digital cellular systems is being prepared. This also will be a dual-mode compatibility standard for both analog and digital units.

10.11 APCO PROJECT 25

APCO Project 25 system standard's major goal is The Common Air Interface for radio transmission of data and digitized voice. Mobile and portables from any manufacturer may be freely combined from any system to work through any APCO Project 25 system. A base line of radio features will be guaranteed from any system to work through any manufacturer's radio.

The Power Behind APCO Project 25

The APCO Project 25 is a joint effort of many groups. The steering committee consisted of four APCO, four NASTD, and three federal government representatives, including the National Telecommunications and Information Agency (NTIA), National Communications Systems (NCS), and Department of Defense (DoD).

The steering committee is encouraging international participation. The United Kingdom, Holland, France, Belgium, Sweden, Norway, Denmark, Australia, Canada, New Zealand, and Austria are some of the countries attending APCO Project 25 meetings.

Many companies, including international corporations, are involved with APCO Project 25: Aware, Inc.; Bendix King, Inc.; Cycomm Corp. (Marconi U.S. Agent); Digital Voice Systems, Inc.; Ericsson G.E.; E.F. Johnson, Inc.; Glenayre Electronics; GTE, Inc.; GEC-Marconi Ltd.; Midland LMR, Inc.; Motorola, Inc.; Racal Ltd.; Raytheon Inc.; SEA, Inc.; and Teletec, Inc.

APCO's vision. APCO Project 25 envisions two generations of radio equipment transmitting data and digitized voice with a migration path from one to the other. The first generation will use 12.5 kHz channels. The second generation narrows the channels to 6.25 kHz.

On January 15, 1993, the APCO Project 25 Steering Committee adopted an interim report on digital radio technical standards. This report described a number of interfaces and stated that each would be examined, one at a time. The steering committee presented some standards for the first interface, The Common Air Interface.

The major technology selections completed are

1. Access method, frequency division multiple
2. Bandwidth, 12.5 kHz
3. Modulation, QPSK-C family
4. Vocoder, improved multiband excitation (IMBE)

5. Channel data rate, 9.6 kb/s
6. Data frame format
7. Encryption, digital encryption standard (DES)

The Common Air Interface. The Common Air Interface ensures that radio equipment which conforms to APCO Project 25 will be interoperable with radio equipment from different manufacturers and compatible with radio systems for different agencies. This allows effective and reliable intra-agency and inter-agency communications in an all-digital mode for voice and data. The IMBE is one of the most important elements of The Common Air Interface.

IMBE vocoder and decoder. Figure 10-5 shows the overall IMBE system. The voice coder converts voice to a digital format. The voice encoder itself operates at 4.4 kb/s, which corresponds to a voice frame of 88 bits for every 20 milliseconds of speech. Voice frames use 56 parity check bits to make the overall size of the encoded voice frame of 144 bits. Golay and RS correction codes protect the IMBE bits.

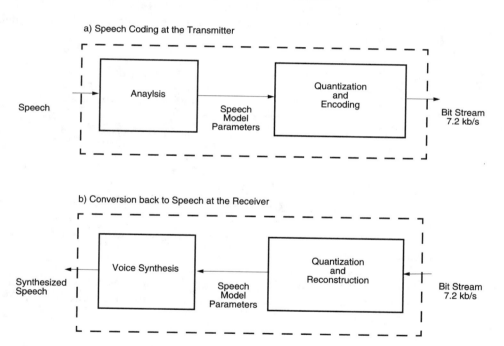

Fig. 10-5 IMBE Speech Encoder and Decoder
By permission of APCO.

Voice and data.   Voice and data messages are sent over the air as packets. In the voice case the packet begins with a header unit and then a superframe of 360 milliseconds and ends with a terminator unit. Data messages are transmitted in a packet whose length is variable.

Trellis, or convolution, codes with code rates of one-half and 3/4 protect packets. Part of the packet is a CRC used to verify the message's accuracy. If the packet is corrupted during transmission, an automatic request (ARQ) is transmitted to repeat the packet.

Modulation.   Two types of transmitter modulation are available for the 12.5-kHz bandwidth. One uses a form of $\pi/4$ DQPSK, which in APCO 25 is designated CQPSK. It uses a partially linear final stage.

The second type is a transmitter that modulates the phase but keeps the carrier amplitude constant, generating a frequency-modulated waveform designated C4FM. The modulation in each case sends 4800 symbols/second (9,600 b/s) with each symbol conveying two bits of information. Table 10-2 shows the relationship between information bits and phase change in CQPSK modulation and information bits and frequency deviation in the C4FM transmitter. The CQPSK uses a partially linear class AB final in the transmitter for four-level phase-shift transmission while the C4FM transmitter uses four-level FM transmission. Both operate on 12.5-kHz channels.

TABLE 10-2 PHASE CHANGE AND FREQUENCY DEVIATION*

Information Bits	CQPSK Phase Change	C4FM Deviation
01	+135 degrees	+1.80 kHz
00	+ 45 degrees	+0.60 kHz
10	- 45 degrees	–0.60 kHz
11	–135 degrees	–1.80 kHz

* By permission of APCO.

Receiver.   The receiver is intended to be compatible with either CQPSK or C4FM transmitters.

Future migration to 6.25-kHz channels.   The transmitter final must be highly linear for future 6.25-kHz channels. The transmitter will use a form of $\pi/4$ DQPSK or C4FM modulation.

10.12 COPING WITH THE CHANGE TO DIGITIZED VOICE

There are three basic strategies to meet the challenge of this great change in radio transmission: backward compatibility, dual mode operation, and software determination.

Backward compatibility. Backward compatibility is the new units' ability to operate with an "old" system infrastructure. The Motorola Astro equipment that can operate with analog voice equipment is one example. The Astro equipment gradually can replace conventional voice equipment. When the radios are all backward compatible they all can be switched to digitized voice.

Dual-mode operation. Dual-mode operation is a system where mobile units may operate in either digitized voice or analog voice. Digitized voice cellular operation is an example. EIA/TIA/IS 54-B cellular system dual-mode mobile-base station forms a compatible standard for cellular mobile telecommunications systems that are incorporating TDMA digital technology in the United States. If a mobile unit is analog voice, the dual-mode system accommodates it. On the other hand if a mobile unit is digitized voice, the system operates in that mode. If a system operates with CDMA digital technology, EIA/TIA/IS 95 dual-mode compatibility standard applies. This standard allows operation in either analog or CDMA digital standard. If a vehicle using TDMA digitized voice technology enters a CDMA system's area, both the mobile and base station switch to the analog voice mode.

Software determination. Software determination is a future method in which specific software defines all the operations of direct-conversion linear translators by employing linear up and down conversion. A software-defined radio automatically can select one of the preprogrammed interoperability options.

REFERENCES

Biglieri, Dariush and Simon McLane. *Introduction to Trellis Coded Modulation with Applications.* New York: MacMillan, 1991.

Brenig, Theodore. "Data Transmission for Mobile Radio," *IEEE Transactions on Vehicular Technology.* August 1978.

Bruckert, E. J. and J. H. Sangster. "The Effects of Fading and Impulse Noise on Digital Transmission Over a Land Mobile Channel," *IEEE Transactions on Vehicular Technology.* November 1974.

Clark, G.C., and J.B. Cain. *Error-Correcting Coding for Digital Communications.* Englewood Cliffs, NJ: Prentice-Hall, Inc., 1983.

Das, J., S.K. Mullick, and P.K. Chatterjee. *Principles of Digital Communications.* New York: John Wiley & Sons, Inc., 1987.

Deasington, R.J. *X.25 Explained, Protocols for Packet Switching Networks.* New York: John Wiley & Sons, Inc., 1986.

Dettmer, Roger. "GSM, European Cellular Goes Digital," *IEE Review.* July/August 1991.

Gates, John. "Packet Radio Combines Computer Radio Technology," *Mobile Radio Technology.* August 1985.

———. "How to Use Packet Techniques to Send Data on Radio Systems," *Mobile Radio Technology.* September 1985.

Glasgal, Ralph. *Techniques in Data Communications.* Dedham, MA: Artech House, Inc., 1982.

Hardwick, J.C. and J.S. Lim. *A 4800 bps Improved Multi-Band Excitation Speech Coder.* Proceedings of IEEE Workshop on Speech Coding for Telecommunications, Vancouver, B.C., Canada, September 5–8, 1989.

Imai, Hideki. "Essentials of Error-Control Techniques." San Diego, CA: Academic Press, Inc., 1990.

Shannon, C.E. "A Mathematical Theory of Communications," *Bell System Technical Journal.* Vol. 27. 1948 628–656.

Document without listed author

APCO 25 Interim Report. Printed and Distributed by the APCO Institute, Inc., South Daytona, FL, 1993.

11

Using Pager Systems

A paging system can be part of a land mobile radio system with a paging encoder at the central dispatching center. The encoder is connected to the base transmitter and the necessary paging tone is sent out on the base transmitter to activate the particular pager receiver desired. The base transmitter still is being used for the conventional two-way radio communications.

Radio common carriers also operate pagers, renting the paging service, including the pagers, for a fee. Sometimes the pagers are sold to the user, who then pays a reduced monthly fee for this paging service.

In addition, there are certain applications for paging-only operations in the special emergency radio services (hospitals, etc.). Here the licensee is not a radio common carrier and two-way communications are not permitted.

This chapter covers special paging definitions, tone sequential paging, and direct telephone paging. Two-tone and five-tone sequential paging are examined. Three different types of digital pagers are discussed: the Nippon Electric Company pager, the POCSAG System, and the Golay Sequential Pager.

11.1 MAKING SENSE OF PAGER TERMINOLOGY

Terms to Know

The following terms are important to understanding paging systems:

Battery-Saving Mode.   The pager strobes on and off looking for the correct preamble. The preamble indicates a transmission for a cluster of pagers. If the pager decoder does not detect its correct preamble, the pager returns to the battery-saving mode where the battery current is at a minimum. If the correct preamble is detected, the pager remains powered up and looks for its proper individual code. Battery saving can extend battery life by approximately three times. However, the pager takes longer to respond in this mode.

Call Rate. The call rate is the number of pagers in a busy hour divided by the number of active pagers.

Camp-On Time. In a dial access paging terminal a second caller gets a "ring back" until the first page is completed. The amount of time the second pager waits while hearing the ring is called camp-on time.

Cap Code. The cap code is the identifying number on the outside of the radio pager. It is related to the tones or digital code that gives the address and other information about this particular pager.

Data Display. There are two types of data display: numeric and alphanumeric. Numeric displays numbers only; alphanumerics shows both letters and numbers.

Dotting. Dotting is a series of zeros and ones at the beginning of a digital pager transmission. Dotting is used for bit synchronization.

Tone Only. In a tone only pager paged parties hear a tone that alerts them to call a phone number that they have previously arranged with the person calling. A message awaits them at this phone number.

Tone and Voice. A voice message follows the alerting tone. In some pagers the recipient has to push a switch to hear the voice message.

11.2 PAGING USING SEQUENTIAL TONES

There are a number of pager tone sequential systems. The two-tone sequential and the five-tone sequential systems are two of the most popular.

Two-Tone Sequential Paging

Figure 11-1 shows some examples of two-tone sequential paging formats with two separate tones, A and B. Figure 11-1a shows the timing for tone only while figure 11-1b is a tone and voice two-tone sequential pager format.

Codes for two-tone sequential pagers. There are a number of codes for two-tone sequential pagers in use. One code uses a cap code of three numbers with a letter prefix: C324 for example. This code determines the filters' frequencies, one for tone A and the other for tone B, in the pager receiver. The encoder in the transmitter must send out these two tones in the timing shown in figure 11-1. To determine frequencies A and B: For

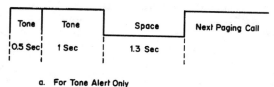

a. For Tone Alert Only

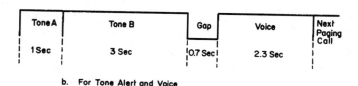

b. For Tone Alert and Voice

Fig. 11-1 Two-Tone Sequential Pager Format

each letter code there are three tone groups of 10 frequencies, or a total of 30 frequencies per letter code. Table 11-1 shows how the tone groups are selected by the letter and first digit of the pager identification code. For a cap code of C324 C selects the tone groups in the two left columns. Then, moving down to the first digit—3—the tone group is 1 for tone A and 2 for tone B. The actual frequencies are then determined from table 11-2. The second digit in the cap code is 2 and this is the tone number in table 11-2. Using tone 2 and tone group 1 gives tone A 369 Hz. Similarly, the last digit in the cap code is 4. Using tone 4 and tone group 2 gives tone B 700 Hz.

There are many variations of this code scheme so this is only one example of the two-tone sequential codes. Two-tone sequential codes can produce 354 unique codes using 60 tones for tone-only pagers.

Five-Tone Sequential Pagers

Five-tone sequential pagers are used primarily for tone-only operation. The five-tone sequential system also is known as decimal digital. Up to 100,000 subscribers theoretically could be served with this tone alert-only system. The decimal digital format takes about 0.2 of a second compared to 2.5 seconds for the two-tone sequential system. Figure 11-2 shows the timing sequence of a five-tone sequential system.

TABLE 11-1 SELECTION OF TONE GROUPS, TWO-TONE SEQUENTIAL PAGER

Code Type:	C		D		E Etc.	
Tone:	A	B	A	B	A	B
First Digit						
1	1	1	1	1	1	1
2	2	2	2	2	2	2
3	1	2	1	2	1	2
4	4	4	1	5	2	1
5	1	4	5	5	1	6
6	1	1	1	1	1	1
7	1	1	1	1	1	1
8	1	1	1	1	1	1
9	1	1	1	1	1	1
0	1	1	1	1	1	1

TABLE 11-2 SELECTION OF TONE FREQUENCIES, TWO-TONE SEQUENTIAL PAGERS

	Tone Group Number			
Tone Number	1	2	3	Etc.
1	350	600	290	
2	369	635	297	
3	390	670	305	
4	411	700	313	
5	434	750	950	
6	460	790	980	
7	480	830	1000	
8	511	880	1035	
		etc.		

Codes for five-tone sequential pagers. In five-tone sequential pagers each digit in the cap code of a pager is assigned a frequency, as illustrated in table 11-3.

For example: the cap code is 2–13756, the tone of the preamble is 880 Hz, and the five sequential tones are 740, 1000, 1590, 1300, and 1400. The encoder at the transmitter sends out these tones to the pager receiver, where an alerting tone would be produced after decoding.

Preamble	Gap	1st Tone	2nd Tone	3rd Tone	4th Tone	5th Tone	Gap	Next Paging
690 MSec.	45 MSec			165 Millisecond			52 MSec	

Fig. 11-2 Five-Tone Sequential Format

TABLE 11-3 PAGER DECIMAL DIGIT TONES, FIVE-TONE SEQUENTIAL

Digit	Tone Frequency (Hz)
0	600
1	740
2	880
3	1000
4	1160
5	1300
6	1400
7	1590
8	1730
9	1870

11.3 PAGING USING DIGITAL TECHNOLOGY

Digital display pagers receive their information in short bursts or packets. Special techniques are used to synchronize digital pagers with their transmitters. In addition special error correction codes are used to minimize errors.

Synchronizing the digital pager. Synchronizing the digital pager with its transmitter involves both dotting and word sync. Dotting is transmitting the reference clock's ones and zeros long enough for the receiver to determine the clock frequency and phase. This allows the pager to bit synchronize with its transmitter. Motorola refers to dotting as comma. Dotting gives the pager bit synchronization and the transmission of a sync word after dotting gives the pager word synchronization. The sync word allows the decoder in the pager to determine the boundaries of each message word. Sometimes more than one sync word is used.

Error correction in digital pagers. Pagers are one-way systems so error-checking systems such as CRC cannot be used. However, there are special error-correcting codes that can correct one-, two-, or three-bit errors. These use code words that consist of a specific number of information bits and a specific number of multiple parity bits. Table 11-4 shows a small segment of a code that can correct one-bit errors.

TABLE 11-4 SEGMENT OF A ONE-BIT CORRECTION CODE

Information Bits	Multiple Parity Bits
1000	110
1001	001
1010	101
1011	010
1100	011

The code word consists of seven bits: four information bits and three parity bits. The code words are constructed so that the minimum bit difference between any two code words is three. This allows a single-bit error to be corrected. A microprocessor compares each received code word with a stored list of correct code words. The microprocessor searches the stored list for a code word that gives the minimum bit difference. If there is a one-bit error in the received code word, the microprocessor selects the unique stored code word that has a one-bit difference. This way a one-bit error can be corrected.

However, a two-bit error is not corrected because there is more than one stored code word that has a two-bit difference with the received code word. However, there are other codes that correct two- or three-bit errors. The $BCH_{32,21}$ code has 32 total bits in a code word, 21 of which are information bits. It can correct two-bit errors. The $Golay_{23,12}$ code has 23 bits in a code word, 12 of which are information bits. It can correct three-bit errors.

Types of Digital Pagers

Nippon Electric Company (NEC) pager. The NEC pager has a data rate of 200 b/s. Figure 11-3 shows the packet format for this system. The sync word is repeated. Each group contains address and display information, and up to four groups can be used. The pager address determines in which group the information is transmitted, which allows the pager to save battery life. Each group has 20 $BCH_{32,21}$ error-correcting code words. Dual-function pagers have both display and tone-alert functions. There is another NEC format that has only one function, tone alert. The page capacity for NEC pagers is 1,048,570 single-function addresses.

Post Office Code Standardization Advisory Group (POCSAG) pager.
The British Post Office developed the POCSAG paging system as a single paging code for the United Kingdom. However, it also is used extensively in the United States. The data rate is 600 b/s. Figure 11-4 shows the POCSAG packet format. After the dotting is trans-

Dotting (Bit Sync)	Sync Word	Sync Word Repeat	Address and Display	Address and Display	Address and Display	Address and Display
			Group 1	Group 2	Group 3	Group 4

Fig. 11-3 NEC Packet for Tone and Message Display

mitted, the addresses and data are sent in batches. The number of batches depends on the number of waiting pages. Each batch starts with a unique sync word followed by eight frames of pager information. The frame consists of two $BCH_{32,21}$ code words: One of the code words is used for an address and the other is used for display characters. Four alert functions are available and the capacity is 2,097,152.

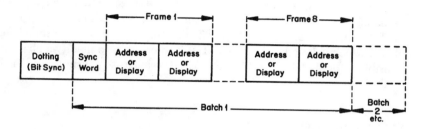

Fig. 11-4 POCSAG Packet

Golay sequential code (GSC) paging. Motorola developed the GSC format, and it uses a $Golay_{23,12}$ error correction code. There are a number of variations depending on the function: tone alert only, tone alert and display, or tone alert with analog voice. Figure 11-5 shows the format for tone alert and display.

Preamble Packet	Sync Packet	Address Packet #1	Display Data		Address Packet #16	Data Display
(Dotting followed by 18 repeats of Preamble Words)	(Dotting followed by two different Sync Words)	(Dotting followed by two Golay Address Words)				

Fig. 11-5 GSC Format

The GSC preamble packet contains 300-Hz dotting for bit synchronization and 18 Golay preamble code words. When bit sync is reached the pager attempts to match a stored preamble with the one received. If a match is found, the pager remains active to receive the sync packet. If not, the pager can go into a battery-saving mode.

The GSC sync packet has dotting followed by two $Golay_{23,12}$ sync words to ensure both bit sync and word sync. For display data up to 16 address packets are sent in batch mode.

Each of the 16 address packets starts with dotting followed by two $Golay_{23,12}$ address words. These address words, W_1 and W_2, also have complements, $\overline{W}_1$ and $\overline{W}_2$ produced by interchanging zeros and ones. Four addresses thus are possible using W_1 and W_2 and their complements.

After the address packet, the display data are sent using a $BCH_{15,7}$ code at 600 b/s. Each display data block can transmit 12 numeric or eight alphanumeric characters.

The GSC paging system's capacity using 10 preamble codes is estimated to be 1,000,000.

Digital paging modulation requirements. Digital paging systems use FSK to modulate a transmitter. This means the transmitter must be capable of direct frequency modulation. As discussed in Chapter 2 most land mobile base transmitters are phase modulated, which indirectly produces FM. It is impractical to modify a phase-modulated base transmitter for digital paging. Digital paging systems use transmitters with direct frequency modulation.

11.4 DIRECT TELEPHONE PAGING

Direct digital paging systems allow the calling party to call the pager directly by using the telephone system. It eliminates the need for a manually dispatched operator system. Sometimes solid-state voice storage is used with the direct telephone paging for voice pagers. This accomplishes two things: It allows the person dialing to hang up as soon as the voice message ends even if the phone is ringing while the paging terminal is busy with another phone. Solid-state voice storage also can compress the speech by eliminating long pauses between words, thus speeding up the radio paging traffic.

REFERENCES

Cromack, Gary T. "Error Detection, Correction for Digital Display Pagers," *Mobile Radio Technology*. October 1983.

Tridgell, R. H. "A Report of Field Tests on POCSAG Pagers," *Telecom*. December 1981.

12

Keeping Your System Operating When the Electrical Power Fails

There are two types of electrical power malfunctions that affect land mobile radio systems. The primary one is a blackout, or a complete breakdown of the commercial power supply. This may last minutes or days. A standby electrical generator can take over to supply the radio system in this emergency. The second malfunction is a momentary dropout. This dropout is measured in milliseconds and it can raise havoc with computers used in land mobile radio systems. The uninterrupted power supply (UPS) is used to prevent any commercial power supply failure from affecting the computers.

This chapter covers the standby electrical system. It compares diesel-driven generators with the gas turbine types. The engine, alternator, and the automatic load transfer control switch are discussed. Maintaining and testing a standby electrical power system are described. An example of a weekly maintenance and test checklist is provided.

A complete solid-state UPS for land mobile radio systems is given in detail, including the rectifier, inverter, static switch, batteries, UPS bypass switch, and the standby diesel generator. A maintenance procedure for weekly, monthly, and annual checks is described.

12.1 STANDBY ELECTRICAL GENERATING SYSTEMS

Figure 12-1 shows one example of a standby electrical generating system. It consists of a diesel engine with a start-stop control system mounted on it.

Using a Standby Electrical System

The engine drives an alternator to supply electricity when the electrical power from the utility fails. Voltage sensors in an automatic-load transfer switch detect when the utility power fails. The automatic-load transfer switch turns on the diesel engine by using a starting battery. Then the switch transfers the load from the utility power lines to the power alternator's output. A voltage regulator keeps the power alternator's voltage output within specified limits. A speed governor on the diesel engine maintains the alternator's frequency output within specifications. When the utility power comes back on, voltage sensors cause the automatic-load transfer to switch the load back to the utility. The start-stop control system then shuts down the diesel engine.

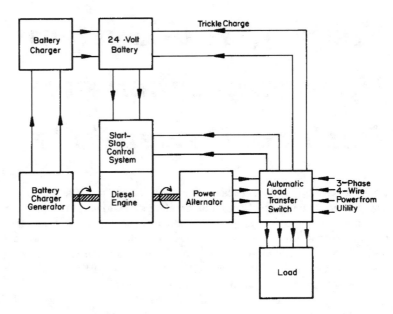

Fig. 12-1 Standby Electrical Generating System

The engine shown in figure 12-1 is a diesel. Gas turbines have been used for standby generators in some land mobile radio systems. Experience shows that the diesel engine has two main advantages for this type of operation. First, the diesel comes up to speed faster than a comparable gas turbine. Second, the wear on a gas turbine depends on the number of starts while the wear on a diesel varies with running time. In backup operation the engine is started frequently to test it, but the running time is very short. This results in more maintenance problems and repairs for the gas turbine. The standby electrical generating system that has been very dependable contains these parts:

Engine. The engine is a four-cycle water-cooled diesel. Using exhaust gases to turbocharge the engine increases the horsepower. The engine has a radiator, fan, and water pump. A hydraulic governor on the engine maintains the alternator frequency from no load to full load within three cycles. The engine has a battery-charging alternator with a solid-state voltage regulator to maintain the 24 volts direct current (V dc) to charge the battery.

A 24-V solenoid shift starter starts the engine. The engine should have a replacement oil filter, oil cooler, and fuel pump.

Engine Instruments. Engine instruments should include an oil pressure gauge, a water temperature gauge, and a battery-charging rate ammeter.

Engine Controls. The engine should start when the start-stop relay closes and stop when the start-stop relay opens. If the plant is not started within 45 seconds, a cranking limiter opens the starting circuit. The controls should include a three-position selector switch with positions run, stop, and remote. Automatic shutdown should be provided for high water temperature, low oil pressure, and overspeed.

Alternator. The alternator should be a brushless four-pole type with a rotating rectifier exciter. The alternator should have a solid-state voltage regulator that keeps the voltage within ±2% of rated voltage from no load to full load. The instantaneous voltage dip when full load and rated power factor are applied to the alternator should be less than 20% of rated voltage. A rheostat should provide a minimum of ±5% voltage adjustment from the rated value.

Alternator Instrument Panel. An instrument panel for the alternator should be provided with the following:

- Running-time meter
- Frequency meter
- Ac voltmeter
- Voltage-adjusting rheostat
- Ac ammeter with phase selector switch

Automatic-load transfer control. The automatic-load transfer control unit performs the following functions:

1. It starts the diesel engine when the utility power fails. After the alternator comes up, it transfers the load from the utility power lines to the alternator.
2. When the utility power is available again, it transfers the load back and stops the diesel engine.
3. It supplies a 24-V trickle charge to maintain the starting batteries in a fully charged state.

Dual standby electrical generators. In many large land mobile radio systems two complete standby electrical generators are used. Each generator has its own complete system, including an automatic-load transfer control. If the first standby electrical generating system does not take over a few seconds after the utility power fails, the second standby generator takes over.

12.2 MAINTAINING AND TESTING AN EMERGENCY POWER SYSTEM

Maintaining and testing an emergency power system is extremely important. An emergency standby system is worthless if it does not operate properly when the utility power fails. It is worth the effort and expense of weekly, semi-annual, and biennial maintenance and test programs.

Weekly Maintenance and Testing Program

This maintenance and testing program has been used over a period of years:

Maintaining batteries and charging circuits.   Inspect the starting battery posts and terminals. When necessary clean the terminals and apply a light coat of Vaseline to them to prevent corrosion.

Inspect the level of electrolyte in the starting batteries. Maintain the level 1/4 in. above the plates, adding water if needed. With a hydrometer test the battery cells for a reading of 1.250 to 1.300. If it is below 1.250, put the charger on high charge until you obtain a reading of 1.275 to 1.300. Then reset the charger to trickle charge.

Check the battery charger instruments' readings. The ammeter should read 0 amperes with an average hydrometer cell reading of 1.275 to 1.300. The ammeter should read above 0 amperes with an average hydrometer cell reading of 1.250. The ammeter should read close to 0 amperes with an average hydrometer cell between 1.250 and 1.275. The voltmeter should read between 13 to 14 V for a 12-V starting system and 26 to 28 V for a 24-V starting system.

Maintaining oil and coolant systems.   Check the level of fuel oil in the main fuel supply tank. Do not allow the level to drop below one-half full. Fill as required.

Check the lubrication oil level in the crankcase. Add lubricating oil as required.

Check the radiator coolant level. Add antifreeze/water solution as required. Inspect the engine system hose connections and look for leaks in the hoses. Inspect the radiator air passage to make sure it is clear. Check the radiator for leaks. Check the generator-bearing lubricant-sight glass gauge for proper oil level as marked. Add lubricating oil as necessary.

Testing the emergency electrical power system weekly.   Put controls on automatic. Operate the test switch if one is provided on the control panel. If a test switch is not provided, open the main circuit breaker. The building electrical load should switch to the emergency electrical power generator after the engine reaches running speed.

Operate the emergency electrical power generator for a minimum of 60 minutes. Then return the test switch or main public utility service switch to the normal position. The building electrical load should switch back to the public utility source, and the emergency electrical power system shuts down automatically.

A Sample Checklist.   The following is a sample of a weekly maintenance and testing checklist for a diesel-fueled standby electrical generator.

Checklist For Weekly Inspection and Test

Generators

Generator Location _____

Date of Test _____ Time_____

Prior to operation:

Battery terminals clean and tight? _____

Battery electrolyte level and condition:

Cell	Electrolyte Level (Proper?)	Hydrometer Reading, Start Test	Hydrometer Reading, after High Charge
1	_____	_____	_____
2	_____	_____	_____
3	_____	_____	_____
4	_____	_____	_____
5	_____	_____	_____
6	_____	_____	_____
7	_____	_____	_____
8	_____	_____	_____

Battery charger put on high charge?_____   Time_____

Battery charger removed from high charge?_____   Time_____

Battery charger reset to trickle charge?_____   Time_____

Indication of battery charger instruments at test start

1. Ammeter reading _____

2. Starting system volts (12 or 24 V) _____

3. Voltmeter reading_____

Controls in "automatic" position? _____

Fuel supply_____ gallons.

Lube oil level: Full? _____ added _____ qts.

Coolant level: Full? _____ added _____ qts.

Coolant system hose connections secure and free of leaks? _____

_____

Radiator air passage clear and radiator free of leaks? _____

_____

Generator bearing lubricant level in sight glass proper? _____

_____

Start OK? _____

Building electrical load switchover OK? _____

End of test record:

    Generator voltage. Phase 1 _____ 2 _____ 3 _____

    Generator current. Phase 1 _____ 2 _____ 3 _____

    Generator frequency _____

    Engine oil pressure _____

    Engine coolant temperature _____

    Hour meter reading_____

    Switch back to building electrical load OK? _____

Semiannual maintenance.   Every six months:

Lubricate fan and water pump bearings.
Lubricate dc charging generator bearings.
Lubricate ac generator main bearings.
Lubricate engine governor.
Lubricate other points and locations specified by manufacturer.
Change engine lubricating oil (crankcase).
Change engine lubricating (crankcase) oil filter.
Change fuel oil filter.
Drain sediment from fuel day tank, main tank, and sediment bowl.
Clean or replace air cleaner.
Clean crankcase breather element.
Check starter and generator brushes.

Biennial maintenance.   Every two years:

Drain and flush engine cooling system in October, every other year.
Replace with fresh antifreeze-water solution.

12.3 CHOOSING AND USING A UPS

Any failure of the electrical power supply, even a momentary fallout, causes a breakdown in computer operations. It takes time to restart computer programs after the power has been restored. In some cases information in the computer can be lost.

A UPS supplies constant power regardless of any power outages. It is used in many land mobile radio dispatching centers, together with standby electrical power generating systems.

UPS for a Land Mobile Radio System

Figure 12-2 shows a solid-state UPS that supplies power to a land mobile radio system. The UPS itself consists of three main sections: the rectifier, the batteries, and an inverter.

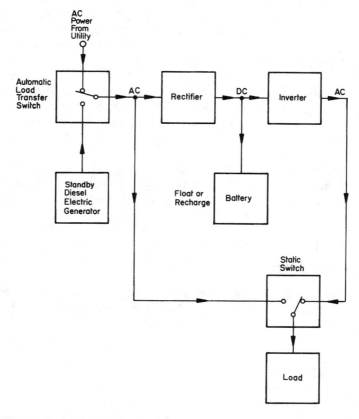

Fig. 12-2 UPS for Land Mobile Radio Systems

UPS Rectifier. This solid-state unit takes the ac power from the utility, rectifies it, regulates it, and filters it. The resulting dc voltage is used for two purposes: to supply the dc voltage for the inverter and to keep the external battery in a charged condition. The rectifier can be operated in either the "float" or "recharge" mode. In the float mode the rectifier provides a small current to maintain the battery in a charged state. It should be noted that the rectifier is the normal dc voltage source for the inverter. If the primary ac input voltage or the rectifier fails, the external battery supplies the necessary dc voltage to the inverter, without interruption, for a limited time. This time depends on the load and the battery size. Some design times are 15 minutes or so.

When the primary ac input voltage is restored, the rectifier again supplies dc voltage to the inverter. This is done in steps over a few seconds, increasing the current gradually until the rectifier is operating at full capacity.

UPS batteries.   The rectifier normally keeps the batteries on a trickle or float charge. After either the ac input or the rectifier fails, the batteries are used and somewhat discharged. After the UPS system is again in operation, the rectifier is placed in the recharge mode and the batteries are recharged. A recharge timer controls the time the rectifier is in the recharge mode. Different types of batteries are available for UPS; the lead calcium storage battery is in extensive use today.[*]

UPS inverter.   The solid-state inverter changes dc to ac. The rectifier provides normal dc voltage input to the inverter. However, if either the input ac power or the rectifier fails, the batteries run the inverter for a limited time. Thus the inverter operation remains uninterrupted.

UPS static switch.   When the inverter output fails or goes below 10% of the normal specifications, the solid-state static switch transfers the load from the utility to the ac line. When the inverter output comes up to within 5% of the nominal voltage, the static switch transfers the load back to the inverter output.

UPS bypass switch.   There is usually a switch that bypasses the UPS for maintenance or in an emergency.

Standby diesel electrical generator.   In the case of a utility power failure, this generator is turned on automatically. During the eight seconds or so before it comes up to speed, the UPS batteries run the inverter and supply ac power.

12.4 UPS MAINTENANCE

Proper preventive UPS maintenance is extremely important. The following is a suggested maintenance schedule based on a number of sources. The actual maintenance depends on the size of the installation and available manpower. The UPS manual also should be consulted for detailed maintenance.

Weekly Maintenance

- Check the water level in all battery cells.

---

[*] The lead calcium battery cells' rated specific gravity varies with the manufacturer.

- Record the readings for one cell per battery (pilot cell) for voltage, specific gravity, and temperature.
- Inspect the battery terminals for signs of corrosion and tightness.
- Measure the voltage across the battery.
- Measure the voltage and current of both the rectifier and inverter.
- Check all system alarms and the operation of the static switch.
- Check the heating, air conditioning, ventilating, and humidity control systems.

Monthly Maintenance

- Give the batteries a booster charge.
- Clean or replace the filters in both the rectifier and inverter.
- Check for signs of fluid leaking from the inverter.

Semiannual Maintenance

- Check all battery cells for voltage, temperature, and specific gravity, and record readings.

Annual Maintenance

- Clean all battery cell surfaces.
- Remove dust from heat sinks in the inverter and rectifier.

REFERENCES

McFatter, Joe H. "Practical Design and Installation Considerations for UPS Systems," *Electrical Consultant*. March/April 1984.

Piper, James. "Keeping Your Uninterruptible Power Supply from Interrupting," *Plant Engineering*. September 20, 1979.

Servos, Gerald H. "Selecting, Installing and Using UPS Systems," *Electrical Consultant*. March/April 1982.

13

Improving Radio Shop Management

This chapter is targeted to large land radio systems that do their own repairs. Planning facilities from the start prevents later problems. Selecting personnel, especially radio technicians, is a very important part of management. Keeping proper records enables management to keep track of radios, check repair efficiency, and maintain a good level of spare parts. The proper management of walkie-talkie batteries saves money and increases the probability of good communications. The question of contracting out or in-house repairs comes up frequently, so a method of comparing the two systems is presented.

13.1 PLANNING THE OVERALL REPAIR FACILITIES

The repair facilities include a number of areas that comprise a complete working environment for maintaining and repairing radio equipment.

Parts of the Radio Shop

Here are the areas that deserve close attention:

Repair Shop Area. The repair shop area should have sufficient space for the technicians, with one working bench per technician. The shop equipment is covered in Section 13.2.

Screen Room. The screen room is in the general repair shop area. It should afford an isolation of at least 70 dB for running tests away from the other equipment.

Inside Area for Vehicle Work. This working area is used for installing radios in vehicles. It should be separate and walled off from the repair shop area. It should have proper ventilation for vehicle fumes.

Office Area. The office area must have space for the shop manager, office clerk, files, telephone, and a computer or computer workstation. It should be adjacent to the repair shop area.

Equipment Storage Area. The equipment storage area must have sufficient storage bins properly marked to keep spare parts and equipment. A desk, chair, and telephone should be provided for the storeroom clerk.

Waiting Area. The small waiting area should have chairs for people who bring in equipment to be repaired or are waiting to pick up repaired radios.

Kitchen-Lunch Area. This area should contain coffee-making equipment, a stove to warm food, a refrigerator, a large table, and chairs.

Conference and Training Room. This room contains a projector, a blackboard, a VCR and TV set, a table, and sufficient chairs.

Rest Rooms. The rest rooms should be adjacent to the shop area.

Shop security. In a large repair shop a great many portable radios and spare parts can "disappear" quickly, adding up to a substantial financial loss, not to mention the embarrassment. The repair shop should be fenced in from the outside, with a locked door and buzzer. Access should be controlled and granted only upon either recognition or proper credentials. The equipment storage area should be fenced in and locked with keys that only the storeroom clerk and shop manager possess.

13.2 EQUIPPING THE RADIO SHOP

The equipment in the radio shop can be divided into bench equipment, shop equipment, and machine shop equipment.

Bench Equipment

The following bench equipment is suggested:

Service monitor. The service monitor is the key instrument in testing and analyzing two-way radio equipment. Because of this, its serviceability is important. Constant use subjects the service monitor to breakdown, which limits the availability of the radio technicians. Some service monitors stand up to continuous use, but others require frequent repairs. Judicious inquiries of other service shops reveals which brands require the least repairs.

The service monitor's major functions and equipment are

- Radio-frequency measurement
- Radio-frequency power measurement
- Speech and tone-coded squelch deviation measurements
- Tone-coded squelch frequency measurement
- SINAD measurement (checks receiver sensitivity)
- Multimode code synthesizer (used to generate all CTCSS tones, paging tones, and digital code squelch signals)
- Signal generator (generates frequencies necessary to troubleshoot radio systems' components)
- Oscilloscope (used with the signal generator)
- Spectrum analyzer (presents a window of the radio-frequency spectrum that is useful in checking for spurious outputs from transmitters, intermodulation interference analysis, and so on)

Additional bench test equipment.   A land mobile radio system also needs

- RF millivoltmeter (a meter reading in millivolts and dBm. The range in millivolts should be from 1 to 3000. The range in dBm should be from –60 to +20. The frequency range should cover 10 kHz to 1 GHz)
- Frequency counter (should be able to measure frequencies from 10 Hz to 1 GHz)
- RF wattmeter (should measure power output from 1 to 50 W over a frequency range from 2 MHz to 1 GHz)
- Portable SINAD meter
- Tone generator (used to generate CTCSS tones and tone bursts)
- Digital multimeter (a Fluke type with an RF probe)
- Volt-ohm meter (a 20,000-Ω/V unit)

Soldering station.   The soldering station should have a vacuum-type solder sucker for desoldering. The soldering iron should be thermostatically controlled with a variable temperature from 525 to 850°. The tip of the soldering iron is grounded.

Bench power.   A 12-V dc power supply adjustable to 40 A should be available for mobile units. Six 115-V ac three-pronged outlets supply power for test equipment.

Large magnifying glass and light.   A magnifying glass and light are necessary for repairing portable radios.

Hand tools.   You need the following hand tools:

Complete sets of regular and Phillips screwdrivers
Complete sets of small and standard nut drivers

Long-nose pliers

Diagonal pliers (cutters)

Hex and spline wrenches

Probers (dental tools)

Vise grips

File standard

Automotive mechanic's tools to remove seats, and so on, to get at radios

Shop Equipment (Not Part of Service Bench)

The following specialized equipment should be available in a large land mobile radio shop depending on circumstances and need:

- Spectrum analyzer. The spectrum analyzer is a more sophisticated instrument than the one in the service monitor. It is useful in analyzing radio interference problems.
- Direction-finding equipment. Direction-finding equipment is useful in locating sources of radio interference.
- Portable battery-charging cycle equipment. Portable battery-charging cycle equipment discharges a battery and then recharges it.
- Telephone control-line tester. A telephone control-line tester measures telephone line loss.
- PROM programmer. The programmer enters data on a new PROM.
- Ultrasonic bath equipment. Ultrasonic bath equipment is used especially in large fire departments to salvage portable radio equipment.
- Engraving equipment. Engraving tools are used to put serial numbers on the outside of radio equipment.

Machine Shop Equipment

The following equipment is recommended for the machine shop:

- Drill press. The drill press should be capable of taking up to 1/2-in. bits.
- Shear and brake. The shear and brake is used to make sheet metal boxes. Brakes should have a 2-ft capability.
- Band saw. The band saw is used to cut piping for mounting antennas.

13.3 SELECTING AND TRAINING PERSONNEL

There are different kinds of personnel necessary to operate an efficient radio shop.

Authorized Personnel

The following positions require certified or specially trained personnel:

Radio technicians. An authorized body must certify the radio technicians as to their competency. The Federal Communications Commission Authorization Act of 1983 gives the FCC the authority to sanction an industry certification program for technicians in lieu of the FCC licenses. Both the National Association for Business and Education Radio (NABER) and APCO have certification programs for radio technicians, and the National Association of Radio and Telecommunications Engineers (NARTE) has a certification program for radio engineers. The FCC no longer requires a commercial radio operator license to operate and maintain two-way land mobile radio equipment. However, the FCC still requires a commercial radio operator license to operate, maintain, or repair radio transmitters in the maritime, aviation, broadcast, or international fixed public radio services. This includes ship radio and radar stations on all types of vessels from small motorboats to large cargo ships, hand-carried portable units used to communicate with ships and coastal stations on marine frequencies, and radios on all aircraft. In addition to being certified radio technicians should have the proper technical education and hands-on experience. Graduation from an appropriate technical school and a minimum of five years repairing radio equipment is required to be a competent technician. In addition to certification, training, and experience a probationary period of at least six months is necessary to determine if a technician is competent.

Radio Technicians' Training. It is extremely important to update radio technicians' training continuously through radio manufacturers' lectures on the latest equipment, video-audio training material sold by radio manufacturers, and magazines dealing with land mobile radio equipment and systems. A library of these magazines should be kept, with periodic cleanup periods to provide room for new issues. Another training source is membership in appropriate organizations where lectures expose people to the latest equipment.

Supertech. Each shop should have at least one "supertech" who, over a number of years, has proven able to handle almost anything that arises. It is important that the person is able to help the younger technicians overcome difficult problems that occur. A "supertech" should pick out one area a month and hold a 20-minute lecture on a particular topic in new technology that applies to land mobile radio. This forces the "supertech" to keep up to date as well as provides the radio technicians new information.

Shop manager. The shop manager should be a certified radio technician with 10 or more years of experience. The shop manager should also be able to schedule workshifts, assign people to various jobs, arrange for equipment and spare parts orders, and oversee all shop functions. In addition the shop manager must maintain good relations with the radio equipment operators. It is important to have one of the senior technicians act as an assistant to the shop manager so he or she can fill in during sick days or vacations.

General-Skills Workers

The following positions do not require specific technical knowledge:

Storeroom clerk. In a large shop the storeroom clerk orders spare parts and stores them in the storeroom. The clerk keeps a running inventory of all spare parts, including spare mobile radios. This information preferably should be entered into a computer.

Office clerk. The office clerk handles the telephone and does the paperwork, typing, and filing. The office clerk should be capable of entering repair work, spare parts, etc., into the computer.

13.4 BETTER MANAGEMENT CONTROL WITH RECORDS AND COMPUTER READOUTS

Computer readouts are useful for keeping track of walkie-talkies, monitoring repair efficiency, and maintaining a spare parts inventory.

Walkie-talkie inventory. Walkie-talkies can be lost or stolen easily because of their small size. Table 13-1 illustrates a readout that presents a sample monthly record of each walkie-talkie's location.

TABLE 13-1 MONTHLY WALKIE-TALKIE INVENTORY READOUT

Month _____			Year _____	
Location: Division, Precinct, etc.	Serial Number	Date in Service	Date Retired	Reason for Retirement[*]
1	C12F1M	01/01/75	01/01/87	A
1	P12F2M	01/01/80		
1	P12F3M	01/01/80		
2	P12F4M	01/01/81		
2	P12F5M	01/01/81	01/01/86	B
2	P12F6M	01/01/81		
Summary:	Walkie-talkies in service		1500	
	Walkie-talkies stolen this month		2	
	Walkie-talkies stolen this year to date		5	

* A, no longer repairable; B, stolen

The walkie-talkie inventory summary in table 13-1 alerts management to problems in safeguarding radios. An increase in stolen walkie-talkies indicates a security problem. Prompt disciplinary action can nip this problem in the bud.

Repair records.   A combination of an individual written card for each radio and a monthly computer readout has been used successfully in tracking repair work. Each walkie-talkie or mobile radio has a repair card filed according to its serial number. The radio technician writes all related test data, repairs, spare parts used, and any comments on this card. Each repair and spare part has a computer symbol listed on the card together with the time for each repair. The radio technician then forwards the signed card to a clerk, who enters the repair code and repair time and the number of repaired radios into a computer. The repair card is then returned to file. At the end of the month a computer readout gives an overall repair picture as shown in table 13-2.

TABLE 13-2 MONTHLY REPAIR ACTIVITY READOUT

Type of Repair	Total Number	Total Hours	Average Time This Month	Average Time Last Month
01	20	40	2	1
02	2	2	1	0.5
03	3	3	1	0.5
04	5	10	2	1
05	7	14	2	1
06	9	18	2	1
07	2	6	3	2
08	3	3	1	1
09	4	4	1	0.5
Summary:	Total number of radios repaired this month	55		
	Total number of radios repaired last month	100		

A sudden drop in radios repaired compared to the preceding month should be investigated. It may be due to a shortage of spare parts, which can be checked against table 13-3. The drop in radio repairs may also be due to a labor slowdown. A comparison of the last two columns in table 13-2 also show a slowdown in productivity. For example, the repair code for 01 represents diagnosis and replacement of potentiometer 156. The average time this month is twice that of last month. If this is consistent for most types of repairs, the readout indicates a labor problem.

TABLE 13-3 MONTHLY SPARE-PARTS INVENTORY READOUT

Part Code Number	On Hand Beginning of Month	On Hand End of Month	10% Level	Monthly Order to Maintain Proper Level
01	80	66	100	34
02	80	80	100	—
03	85	85	100	—
04	80	20	100	130
05	70	70	100	—

This readout can also indicate a problem with specific parts. For example, in table 13-2, 20 01 repair codes represent 20 failures of potentiometer 156. Immediately notify the manufacturer that this part number has an excessive failure rate.

Spare parts inventory control. There are two extremes in maintaining spare parts. One is having too many on hand, which ties up money and takes up too much space. Excessive spare parts also can result in an unpleasant audit by an appropriate authority. The other extreme is not having enough spare parts on hand, which results in having to wait for repairs. This can lead to some unpleasant remarks from the equipment users. The goal is to keep the spare parts level at approximately 10% of total parts in use. Temper these limits for a particular item by experience. There are items that fail frequently, such as volume potentiometers in walkie-talkies. Here the spare-parts inventory should be approximately 15% of total parts in use. On the other hand other parts may rarely fail, and the spare-parts level should be 5%. Document these exceptions to justify the levels to outside authorities.

13.5 SAVING MONEY ON WALKIE-TALKIE BATTERIES

Portable two-way radios use nickel-cadmium batteries that exhibit "memory." If they are given a relatively small use and then charged, they do not discharge all the way. Battery-cycling equipment has been developed to prevent this and use portable radios' nickel-cadmium batteries in the most efficient manner. With this a battery management program can be set up in the radio shop to obtain optimal usage of these batteries, to save money, and to make the portable two-way radio operate more effectively.

Charge-discharge cycling equipment. Charge-discharge cycling equipment charges and then discharges the battery through a resistive load and reads the percentage of the battery's capacity. The discharge should not go below 1 V per cell. The percentage reading indicates the battery's condition: a reading of 50 or lower indicates a poor battery that should be replaced; a reading between 50 and 75 means the battery is in marginal condition and should be cycled three times to increase its capacity; and a reading between 75 and 100 indicates a good battery capacity.

Cycling new batteries.   Fully charge new batteries in accordance with the manufacturer's instructions. Cycle the battery through a charge-discharge cycle and read the percentage of capacity. Some battery manufacturers consider 80% and above acceptable for new batteries. Return batteries under the warranty that are below the manufacturer's rating. This practice weeds out bad batteries in a new shipment. If the number of rejects is large, alert the manufacturer to avoid further trouble.

Cycling with operational batteries.   In a relatively small radio system the batteries can be passed through the charge-discharge cycle every other month. Those that read below 50% should be replaced. Those that read 75% of capacity or over should be passed through. Those between 50 to 75% should be recycled up to three times to get a reading of 75% or over.

Make a record of all readings. Each battery's readings per time it is cycled should be clustered. If the readings are far apart, the battery should be cycled more than once every other month. However in a very large radio system—over 1000 portable radios—cycling every battery every other month may not be practical. One solution may be to cycle the battery when the radio is sent in for repairs.

Cold weather precautions.   Charging the battery right after it comes in from the cold may damage it. The battery should be allowed to come up to room temperature before it is charged either in a regular charger or in a charge-discharge cycling device.

13.6 CONTRACTING VERSUS IN-HOUSE REPAIRS

The question of whether to contract radio repairs or do them in-house comes up from time to time. Two situations are common: when radio repairs in an organization always have been contracted out and it decides to investigate the feasibility of establishing an in-house repair facility, and when an internal radio repair organization exists, but it decides to contract out a portion of the work rather than pay overtime.

Factors to Consider

In both cases there are a number of factors to consider. One, of course, is cost. Cost comparisons are discussed here for each of the situations using the annual cost analysis method. Which method and technique used is important.

The expenses are only for the illustration purposes and may be quite different in actual practice. One expense is the cost of money, which is assumed to be 10% compounded annually in the example. The test equipment life is assumed to be 15 years with negligible salvage value.

Cost analysis. In this example a municipal facility, a fire department, contracts out its radio repair work. It is proposed that the fire department perform its own radio repair work in part of a building the municipality owns. An analysis of this proposal must include a cost analysis that compares the cost of repairing radios in-house with the cost of contracting. The proposed facility requires some building conversion, and the facility also needs tools and test equipment. Thirteen employees are required for the new repair facility. The initial and annual costs of maintaining and repairing radios in-house are calculated. The initial costs are converted to annual costs.

Fixed initial costs of proposed in-house facility:

Building conversion costs	$100,000
Tools and test equipment	300,000
Total fixed initial cost	$400,000

Annual costs of proposed in-house facility:

10 Radio mechanics	$300,000
1 Manager	35,000
1 Clerk	20,000
1 Storeroom clerk	25,000
Repair parts	30,000
Heat	10,000
Electricity	10,000
	$430,000

Converting Initial Costs to Annual Costs. For comparison purposes, it is necessary to convert the initial costs to an annual uniform series of payments. This is accomplished by the formula

$$R = P\left[\frac{i(1+i)^n}{(1+i)^n - 1}\right]$$

where

R = annual uniform series of payments
P = initial cost
i = interest rate, in this case 10% compounded annually
n = number of compounded periods, in this case 15 years

Substituting values yields

$$R = 400,000\left[\frac{0.10(1+0.10)^{15}}{(1+0.10)^{15}-1}\right]$$

The term in brackets is known as the capital recovery factor and can be evaluated by a hand calculator: $R = 400,000(0.13147) = \$52,588$.

Total Annual Cost of Proposed In-House Facility. The total annual cost is the sum of the initial investment's annual cost equivalent and the annual costs, or $52,588 + 430,000 = \$482,588$.

Annual Cost of Contracting. The cost varies with each contractor. To illustrate this analysis method, assume a figure based on the estimate that one contractor's technician can generate $5000 per month or $60,000 per year. For 10 technicians this is an annual cost of $600,000.

Comparing the Annual Costs. The annual cost comparison is between $482,588 for the in-house facility and $600,000 for contracting. The in-house facility would be chosen on this cost basis.

Break-Even Point. The break-even point is the number of years, n, it would take to get back the extra investment in the in-house facility. To evaluate that time span:

$$\text{initial cost } A + \text{annual cost } A(n) = \text{initial cost } B + \text{annual cost } B(n)$$

Substituting the values obtained in this example gives

$$400,000 + 430,000(n) = 0 + 600,000(n); \quad n = 2.35$$

The break-even point is between two and three years.

Other comparison factors. Cost is only one factor in comparing in-house repairs to contracting out. Factors such as immediate service availability, quality of work, flexibility, and specialized personnel availability also should be considered.

The quality of a contractor's work cannot be judged until feedback from the actual users of the equipment is received. There may be complaints to iron out.

It is not always possible to get 24-hour immediate service from an outside contractor. Other commitments can keep an outside contractor from giving priority when it is needed.

Unexpected situations may arise that demand a number of people concentrate on an immediate problem. In-house personnel can sometimes handle this better.

Sometimes in-house personnel cannot properly maintain specialized associated computer equipment. In this case contracting out may be the best method. Each situation must be considered on its own merits.

Contracting Versus Employee Overtime. Overtime pay is another situation that comes up from time to time. Here a direct cost comparison can be made between the contract cost and the in-house employees' overtime salaries. In-house overhead costs remain the same and need not be considered. However there are sometimes legal constraints against overtime that may require outside contracts even though that cost is greater. On the other hand contracting out on a low-bid basis may result in inferior work compared to that from in-house personnel overtime. Again the various factors discussed in this chapter must be weighed together with the cost factor.

REFERENCES

Manko, Howard H. *Soldering Handbook for Printed Circuits and Surface Mounting.* New York: Van Nostrand Reinhold Company, Inc., 1987.

Rensabene, Saverio, and James W. Gould. "Unwanted Memory Spooks Nickel-Cadmium Cells," *IEEE Spectrum.* September 1976.

14

Minimizing Lightning Damage to Your System

Modern land mobile radio systems are becoming much more vulnerable to damage from lightning than ever before. Small currents induced by lightning strokes at a distance can damage integrated circuits. These integrated circuits are everywhere in land mobile radio systems, including computers and their peripherals. This chapter discussed proper grounding techniques and describes ways to decrease lightning damage.

14.1 UNDERSTANDING THE EFFECTS OF LIGHTNING ON LAND MOBILE RADIO SYSTEMS

When a lightning storm occurred in the Brooklyn borough in New York City recently, teleprinters in 15 firehouses broke down because integrated circuitry failed. In the old days of mechanical relays and vacuum tubes, the relatively small currents induced in firehouses some distance from the lightning stroke would have had no effect. Today not only the radio base station is vulnerable—a lightning stroke some distance away can damage any computer equipment and associated peripheral equipment severely. Therefore, it is necessary to have a general understanding of lightning even on an elementary level to minimize damage. It is generally agreed, however, that lightning effects cannot be eliminated completely.

Nature of a lightning stroke.   In most lightning strokes the lower parts of rain and snow clouds from 5000 to 15,000 ft above the earth are negatively charged. These negative charges induce positive charges in the earth below. The cloud-to-ground potential before a stroke is 10^8 to 10^9 V. Air breakdown requires 10^6 V/m. Thus the potential between cloud and ground is not sufficient for a direct lightning strike between these entities. However the wind above the ground can blow positive charges through the air. A negative leader drawn from the cloud travels down toward the earth in jagged steps as it is attracted to randomly placed positive charges in the atmosphere. When the negative leader falls below 600 ft, it can draw sparks from ground objects and structures that come up to meet the leader. When the electric path is complete, a massive return-stroke current results. A number of strokes is called a flash. Usually a flash is no more than five strokes of declining peak amplitude. The duration of a flash is less than 0.5 second with an interval of between 0.1 and 0.25 second between stroke. A lightning stroke's rise time typically is 2 μs. The delay to half of peak amplitude is typically 40 μs. A stroke's median peak current is 18 kA. However a small number of lightning strokes can reach over 100 kA peak. Stroke currents as high as 160 kA have been measured.

Height of structures and lightning. For heights above 25 m (82 ft) the frequency of lightning strokes is proportional to the square of the structures' height.

Potential on a line conductor. Lightning strokes to ground can cause current to flow, which can set up a high potential on the wire. This potential V can be calculated

$$V = I(Z) = I \sqrt{L/C}$$

where

I is the current traveling along the line and Z is the surge impedance of the line = $\sqrt{L/C}$.

Lightning's damages. Lightning can shatter nonconductors and burn or vaporize conductors. Lightning can crush hollow conductors or loops in cavities because of heating and magnetic effects. Pressure inside an equipment building due to lightning sparks can blow the building apart. Current surges due to lightning can destroy components in radio, computer, and other electronic equipment. These currents on wires can destroy integrated circuits some distance from the lightning stroke. And lightning can kill personnel inside buildings.

Minimizing Lightning Damage

There are a number of basic principles that are used as general guides in preventing major damage in land mobile radio systems.

Path of least impedance. The electric current a lightning stroke causes behaves like all electric currents. Given two paths to ground most of the current takes the path with the smallest impedance. Figure 14-1 illustrates this principle. In the undesired path coiling up wire can increase the inductance. The lightning stroke has a typical rise time of 2 μs. This fast rise time means that a small inductance has a high impedance to lightning strokes. Using the smallest wire size compatible with the wire use can increase the resistance in the undesired path. In the desired path the resistance and inductance should be held to a minimum. The wire size used for grounding should be sufficient to carry large currents to ground with little resistance. The radius of curvature of conductors carrying lightning currents to ground should be large to minimize the inductance.

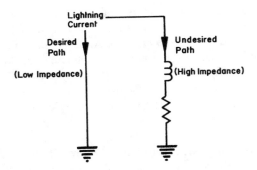

Fig. 14-1 Taking the Path of Least Impedance

Properly grounding the antenna tower.   Above 82 ft the number of lightning strokes increases with the square of the tower's height. Most antenna towers are generally above 200 ft, so it is essential that the tower be grounded properly. This involves the soil's resistivity, the number of ground rods, periodically tightening the structure's bolts to keep the resistance down, and other factors.

Connecting all grounds together.   The antenna tower ground, equipment grounds, telephone company grounds, and building grounds all must be tied together to prevent very high potentials being built from one ground terminal to another during a lightning storm.

Using the proper surge protector.   Surge protectors are used to minimize lightning damage. Lightning strokes' fast rise time must be considered when selecting surge protectors. Another essential factor is the surge protector's ability to handle the large currents lightning produces. A third factor is certain types of suppressors' clamping voltage. These suppressors prevent the voltage across a pair of lines from rising above a certain value called the clamping voltage. The clamping voltage must be above the normal voltage on the line.

14.2 PROTECTING YOUR ANTENNA TOWER

The antenna tower should afford a low-impedance path to ground compared to the impedance path through the radio equipment. The antenna should be the direct dc grounding type, not the type that uses a gap for lightning protection. (A direct dc grounding type of antenna offers a lower-impedance path to ground for lightning.)

Figure 14-2 shows an example of the direct dc grounding type of antenna. In addition there are collinear and other antennas that have a dc connection to the tower. Consult antenna catalogues to select a base station with dc grounding. This is especially important in lightning-prone areas.

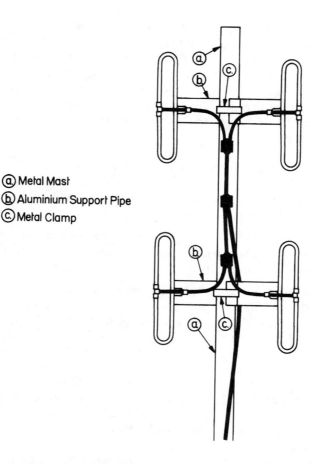

@ Metal Mast
ⓑ Aluminium Support Pipe
ⓒ Metal Clamp

Fig. 14-2 Antenna with dc Ground Through Support Pipe for Lightning Protection
By permission of the Antenna Specialist Company.

    The coaxial cable from the antenna is bonded to the tower at the bottom of the tower. A conduit should be used inside the building at the point where the coaxial cable enters a building. At the point of entry into the building both the conduit and the outer shield of the coaxial cable should go to the same ground point. The coaxial cable leaves the conduit where the conduit comes to the vicinity of station cabinet equipment. Surplus semiflex coaxial cable should be coiled up before it enters the cabinet to give a higher-impedance path to the radio equipment. At the point where the coaxial cable enters the cabinet the outer shield of the coaxial cable should be grounded. This ground point should be common to the cabinet ground. Note that the coaxial coils at the cabinet's entrance can flash over in self-supporting tower installations.

A coaxial impulse suppressor should be used in addition to the coaxial coils. If the equipment building is closer than 5 ft to the antenna tower, the coaxial impulse suppressor is mounted on a grounded tower leg. If the distance is greater than 5 ft, the suppressor should be mounted on a grounded bulkhead panel. If a grounded bulkhead is not used, the coaxial impulse suppressor is mounted on a grounded input/output panel of the equipment cabinet. The next item to consider is properly grounding the antenna tower so that lightning striking the tower has a lower-impedance path to the ground compared to the path to the radio equipment.

Grounding the antenna tower.   A good tower ground should have a very low impedance. This means a low resistance (approximately 10 Ω) and a low inductance. Using a conductor with a large cross section and a minimum bending radius to the ground electrode can minimize a low inductance.

Types of Soil.   Soil's resistivity in Ω/cm varies greatly with the type of soil. Clay runs from 2500 to 7000. Sand's resistivity is from 100,000 to 300,000 Ω/cm. Wet soil considerably reduces soil's resistivity.

Using Grounding Rods.   The minimum standard for grounding rods according to EIA RS-222-C is two 5/8-in. rods (copper-clad steel) 8 ft long on opposite sides of a self-supporting tower. One grounding rod should be used per leg for self-supporting towers greater than 5 ft per side. Each rod is a minimum of 1 1/2 ft below grade and 2 ft minimum from the foundation. There should be 6 ft minimum separation between rods. A No. 6 solid copper wire connects the nearest leg of metal base to the grounding rod with no sharp bends in the wire. The No. 6 copper wire should be as vertical as possible. If bends in the No. 6 conductor must be placed above the earth, the minimum radius of the bend should be 8 in.

Increasing the number of ground rods decreases the resistance, as does increasing the separation between grounding rods. Some cautions: Do not use stranded wire or braid in grounding, and copper and galvanized steel should not be placed in direct contact with each other. They may be used in conjunction with a bronze fitting that acts as a buffer.

Radial Grounding.   In some situations, as on a mountaintop site, it is difficult to drive in grounding rods. Here radial wires can be placed on the ground or just underneath the surface. In 1000-Ω/m soil, eight No. 10 radial wires, each 50 m long, would provide a ground resistance of 13 Ω.

Combining Ground Rods and Radials.   Using both ground rods and radials results in a lower resistance than using either method alone. This can be done by adding a grounding rod driven into the earth at the end of each of the radials, or along the length, as long as they are separated sufficiently.

Grounding for guyed towers.   Grounding a guyed tower involves the same steps as a self-supporting tower, with the addition of a similar ground at each nongrounded guy anchor. The ground rod at the guy anchors should be bonded to each guy on the tower side of the turnbuckles to avoid putting an unusual strain on the turnbuckles.

Grounding a tower on a building with a lightning protection system.
When a building has a lightning protection system the antenna tower must be bonded to the main conductor of the lightning protection system. The bond must be the same size as the main conductor of the lightning protection system.

Grounding a tower on a building without a lightning protection system.
One suggestion for a self-supporting tower on an unprotected building is to use two main-sized conductors.* Each conductor is bonded to the tower and then run down opposite sides of the building to independent protection ground rods. At least one of the ground rods is bonded to the building water piping. The ac neutral is bonded to the same water piping system.

Protecting side-mounted antennas.
Special equipment is required to protect side-mounted antennas mounted over 50 ft up the tower from direct strokes. Two horizontal lightning rods, one just below and one just above the side-mounted antenna, afford a good measure of protection without affecting system performance. They should extend horizontally at least 6 in. beyond the antenna.

Tower lighting cable.
The tower lighting cable supplies electricity to the tower lights. The cable should be a twisted pair in a grounded metal conduit to produce a series inductance effect in the cable, which offers the lightning stroke a higher-impedance path.

14.3 PROPERLY GROUNDING YOUR SYSTEM AND MEASURING GROUND RESISTANCE

It is extremely important to properly ground the equipment building, power lines, and telephone lines as well as the antenna tower.

The CADWELD ground connection.
Many grounding specifications now require the CADWELD Process† for joining cables to ground rods and for general use in bonding components of a ground system.

The CADWELD Process is a method of making copper to copper or copper to steel electrical connections that do not require an outside source of heat or power.

Powdered copper oxide and aluminum are put into a graphic crucible and ignited by a flint igniter. The aluminum reduces the copper oxide, which produces molten copper that flows over conductors, melting and welding them together.

Common system ground.
A lightning stroke of 100 kA and a ground system resistance of 10 Ω produces a peak potential of 1 million volts. If the nearby equipment

---

* This is the type of conductor used as the main conductor in building lightning protection systems.
† This is a registered product of ERICO Products, Inc., 34600 Solon Road, Solon (Cleveland), Ohio 44139.

building grounds, power lines, and telephone lines are separate, a great deal of damage can result. All the grounds must be interconnected as a grid by using a separate line to connect each ground to every other ground or by using a closed loop completely around the equipment building.

Principles of measuring ground resistance.   There are several commercial instruments that measure ground resistance. These usually have detailed operating instructions, so this chapter only covers some of the basic principles.

Figure 14-3 shows a procedure for measuring an antenna tower's ground resistance. The ground resistance is calculated by dividing the voltage reading by the current reading. The signal source should not be at a frequency of 60 Hz or its harmonics to eliminate the effects of stray power system currents.

The position of the two test probes in Figure 14-3 is very important to obtain correct readings. Test probes 1 and 2 should be in line with the antenna tower. Distance B should

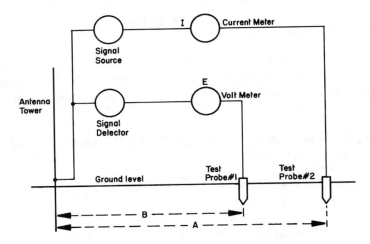

Fig. 14-3 Measuring Antenna Tower Ground Resistance

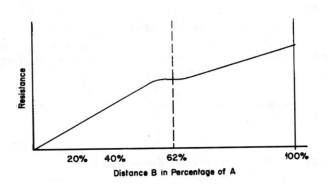

Fig. 14-4 Ground Resistance Versus Distance B

be 62% of distance A. Distance A must be great enough to obtain correct readings. If distance A is great enough, measuring the earth resistance versus distance B produces a curve that flattens out (point of inflection) at the 62% distance, as Figure 14-4 shows. If the curve in Figure 14-4 were a straight line, it would indicate that test probe 2 is too close to the antenna tower. After probe 2 is moved out, probe 1 must be moved so it is 62% of the distance from the antenna tower to test probe 2.

14.4 SURGE PROTECTORS AND THEIR CHARACTERISTICS

To use surge protectors properly to minimize lightning damage it is necessary to know these devices' operating characteristics. Surge protectors are divided into two general types: primary and secondary.

Primary and Secondary Transient Suppressors

The primary device can handle large currents but has relatively slow response times. The primary device is usually a gas discharge tube. In contrast the secondary type handles smaller voltages and current surges but has a much faster response. There are a number of secondary types, including varistors and PN silicon voltage suppressors (a specialized type of zener diode).

Gas Discharge Tube. The gas discharge tube contains an inert gas and has two or more electrodes. When its voltage rating is exceeded, the tube becomes conductive. The response time is relatively slow, but it can handle large currents. After discharge it can be used over again. Figure 14-5 shows the symbols for a two-terminal gas discharge tube and a three-terminal gas discharge tube together with other suppressor types.

Metal Oxide Varistor (MOV). A varistor is a nonlinear resistor. The resistance changes with the voltage applied. The response time is approximately 10^{-9} second. Most varistors are made of zinc oxide and other metal oxides. Varistors can survive thousands of surge current amperes. Some varistors are built into a three-prong plug. Others are mounted on leads.

PN silicon voltage suppressors. Some PN silicon voltage suppressors have a fast response time on the order of 10^{-12} second. These units are used in secondary applications.

Combining primary and secondary voltage suppressors.
Figure 14-6 illustrates a combination of suppressors that uses a primary voltage suppressor (a gas discharge tube) and an MOV secondary voltage suppressor. The MOV has a fast response time but does not handle as large currents as the primary voltage suppressor. The gas discharge tube handles large currents, although its response time is slower. Together these two devices form a unit that responds quickly to lightning surges and handles large currents without being destroyed in the process. The small resistors R_1 and R_2 are used to decouple the MOV from the gas tube, allowing the gas tube to "fire" without having the MOV's voltage set too high. Inductances may be substituted for R_1 and R_2.

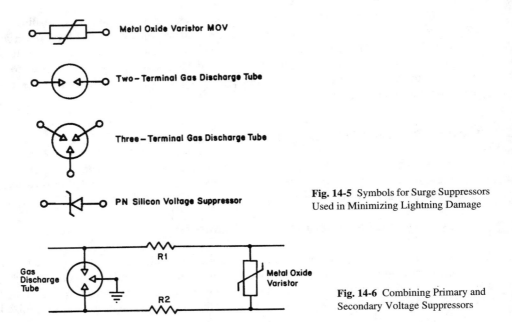

Fig. 14-5 Symbols for Surge Suppressors Used in Minimizing Lightning Damage

Fig. 14-6 Combining Primary and Secondary Voltage Suppressors

14.5 USING SURGE PROTECTORS TO CONTROL LIGHTNING DAMAGE

There are three general areas that should be protected by surge-limiting devices: power facilities, telephone lines, and equipment such as radios, computers, and computer peripherals.

Power lines. The main power-line input and the equipment cabinet power lines require special treatment.

Main Power-Line Entrance. Metal oxide varistors can be used for main power-line entrance with ratings of 50 kA, 420 V clamping at 200 ampere (A), and 5- to 10-nanosecond response time. The GE V151BA60 is an example. The protector is located as close as possible to the power-line entrance in an electrical knockout box.

Equipment Cabinet Power Lines. With equipment cabinet power lines use three power-prong plugs with three connections: high, neutral, and ground. Introduce inductance in the high and neutral lines by using an isolation transformer. Use an MOV from high to neutral across the power lines. The MOV should match the lead type installed inside the equipment cabinet.

Equipment Cabinet with an Antenna Connection. When the equipment cabinet has an antenna connection, use a metal oxide varistor as with power lines. In addition use a two-terminal gas discharge tube from neutral to ground at the equipment cabinet. The gas discharge tube handles the large amount of current due to a lightning stroke coming from the antenna.

Telephone lines. The telephone company installs a carbon block unit on the telephone lines to protect against lightning. Install a three-terminal gas tube adjacent to the carbon block. Two of the terminals go to each of the telephone lines and the third terminal goes to ground. If a good ground is not available, use a two-terminal gas tube protector across the telephone lines. The gas tube protects the carbon block. The carbon block acts as a back-up in case the gas tube fails.

Other signal lines. There may be lines transmitting current from fire alarm boxes or other equipment. MOVs should be used across these lines. The regular voltage across these lines may vary considerably. However it is important to know the highest voltage value so that an MOV with a higher clamping voltage is selected.

REFERENCES

Block, Roger. *The Grounds for Lightning and EMP Protection.* 2nd e. Polyphaser Corp., P. O. Box 9000, Minden, NV 89423–9000, 1993.

————— . "Lightning Protection Guide for Radio Communications." Polyphaser Corp., P. O. Box 9000, Minden, NV 89423–9000.

————— . "Grounding Standard Guides Lightning Protection Installation," *Mobile Radio Technology.* September 1986.

Bodle, David. "Electrical Protection Guide for Land-Based Radio Facilities." Joslyn Manufacturing and Supply Co., Box 817, Goleta, CA 93017, 1971.

Curdts, E. B. "Some of the Fundamental Aspects of Ground Resistance Measurements," Pap. No. 58–106, *AIEE Transactions.* Vol. 77. 1958.

Gross, Al. "Practical Notes on Surge Protection," *Proceedings of the Radio Club of America.* November 1987.

Guthrie, A. K. "Living with Lightning." Seminar notes, GE Mobile Radio technical training, ECP–826A, 1979.

Keiser, Bernhard E. "Lightning Control, Syllabus for a Training Course." Don White Consultants, Inc., International Training Centre, State Route 625, P. O. Box D, Gainesville, VA 22065, 1980.

Publications without listed authors

About Lightning. Decibel Products, Inc., 3184 Quebec Street, Dallas, TX 75247 (not dated).

Getting Down to Earth—A Manual on Earth-Resistance Testing for the Practical Man. Biddle Instruments, 510 Township Line Road, Blue Bell, PA 19422, 1982.

Lightning EMP and Grounding Solutions Catalog '93. Polyphaser Corp., P. O. Box 9000, Minden, NV 89423–9000, 1993.

Lightning Protection Code. National Fire Protection Association, Inc., 470 Atlantic Avenue, Boston, MA 02210, 1986.

Transient Voltage Suppression Manual. General Electric Co., Semiconductor Products Department, 1988. Available from electronic component distributors.

15

Understanding Land Mobile Radio Propagation

It is important to understand radio propagation for land mobile radio for a number of reasons. Multipath propagation between a vehicle and a base station prevents 100% reliable communications throughout a service area. A probability figure of reliability can be calculated from propagation formulas for a given service area. Propagation prediction can also determine the base transmitter antenna height and power so that propagation is limited primarily to the service area.

In this chapter we examine the different types of propagation: free-space, plane-earth, and scattering. A method to determine when to use free-space or plane-earth propagation formulas is presented. Different propagation models that predict service coverage are examined. A method of calculating probability degradation for different levels of reliability for your radio system is presented, along with a method to measure your system's service coverage. Digitized voice's effect on reception is presented graphically.

15.1 EXAMINING THE GENERAL TYPES OF PROPAGATION

Free-Space Propagation

Free-space propagation takes place between a transmitting and a receiving antenna in empty space with no interfering bodies between. There is no ground reflection nor diffraction losses. Free-space attenuation occurs during the first few miles of path. Free-space loss is 6 dB per octave of distance (doubling the distance). The formula for free-space attenuation using a dipole antenna for both antenna and receiver is

$$\text{attenuation (dB)} = 32.6 + 20 \log F + 20 \log d$$

where

F is the frequency in megahertz and d is the distance in miles between the transmitting and receiving antennas. Diffraction losses as the radio wave passes over an obstacle is one limitation to free-space propagation.

Diffraction losses and Fresnel zones. A wavefront expands as it travels, causing reflections and phase changes as it passes over an obstacle. This results in a diffraction loss of signal. The Fresnel phenomenon occurs in zones. The accepted added clearance to an obstacle is 0.6 of the first Fresnel zone. The added clearance can be calculated

$$\text{added clearance (ft)} = 0.6(2280) \left[\frac{d_1 d_2}{L(d_1 + d_2)} \right]^{1/2}$$

where

d = distance from transmitter antenna to path obstacle, miles
d_2 = distance from receiver antenna to obstacle, miles
L = wavelength, ft

Range of free-space propagation. Pure free-space propagation extends from the beginning of the far field[*] of the transmitting antenna to a point where the direct wave between the two antennas fails to clear the first obstruction by at least 0.6 of the first Fresnel zone. At this point additional losses occur. For 160 MHz free-space propagation over a clear path extends from 60 ft to 5 miles from the transmitting antenna for antenna heights of 110 ft.

Plane-Earth Propagation

Plane-earth propagation is the multipath propagation due basically to the reflection from a smooth earth. Plane-earth attenuation is twice that of free-space attenuation with distance. That is, plane-earth propagation is attenuated 12 dB per octave of distance compared to the 6 dB per octave of distance for free-space. The formula for plane-earth attenuation in decibels is

$$\text{attenuation (dB)} = 144.6 + 40 \log d - 20 \log (H_T H_R)$$

where

d = distance between transmitting and receiver dipole antennas, miles
H_T = height of the transmitting antenna, ft
H_R = height of the receiving antenna, ft

---

[*] The far field is a distance approximately 10 times greater than the wavelength from the antenna.

Differentiating between plane-earth and free-space propagation.   To differentiate approximately between free-space and plane-earth propagation:[*]

$$d = (2.3 \times 10^{-6})F(H_T H_R)$$

where

d	=	distance, miles
F	=	frequency, MHz
H_T	=	height of transmitting antenna, ft
H_R	=	height of receiver antenna, ft

When the path distance is less than d, free-space propagation is used. When the path distance exceeds d, plane-earth propagation is used. For example:

F	=	160
H_T	=	$H_R = 110$ ft
d	=	$(2.3 \times 10^{-6})(160)(110)(110) = 4.5$ miles

This means plane-earth propagation occurs at greater than 4.5 miles. An actual measurement[†] shows that plane-earth propagation (12 dB per octave of distance) starts at about 5 miles for these conditions.

Plane-earth propagation and the Earth's curvature.   The bulge due to the Earth's curvature is an obstacle to plane-earth propagation. Radio waves are refracted or bent as they go from dense to less dense air. This results in an effective radius of the earth, K, which is greater than R, the physical radius of the earth. For many calculations $K = 4/3R$ is used. In practice K depends on the refraction caused by many weather and geographical factors. The radio horizon for a base-station antenna over a smooth earth can be calculated from the formula $H_R = (2h)^{1/2}$. H_R is the radio horizon in miles and h is the height of a base transmitter in feet. If the base transmitter is 200 ft above the ground, the radio horizon is 20 miles. The earth's curvature causes diffraction losses, which increase with distance and add to the plane-earth losses.

Scatter Propagation

Scatter propagation occurs at distances greater than 50 miles. It has an attenuation slope of 20 dB per octave, increasing to beyond 30 dB per octave at distances in the hundreds of miles for VHF high-band.

---

[*] From "Engineering Considerations for Frequency Coordination" by Robert L. Gottschalk (see References).
[†] "VHF and UHF Propagation," *GE Data File Bulletin No. 1003-1*, July 1962.

15.2 USING DIFFERENT PROPAGATION MODELS*

Different propagation models that incorporate various losses have been used. These models are attempts to practically predict propagation losses. Some of these are detailed here (most of them are still in effect):

Bullington model.   Kenneth Bullington published his Bullington model in the October 1947 issue of the *Proceedings of the IRE*. Bullington modified the basic free-space and plane earth models by including diffraction losses from both the Earth's curvature and objects such as hills. A number of nomographs give these diffraction losses. Figure 15-1 shows one of these with some changes in the frequencies. The distance d_1 is less than or equal to d_2.

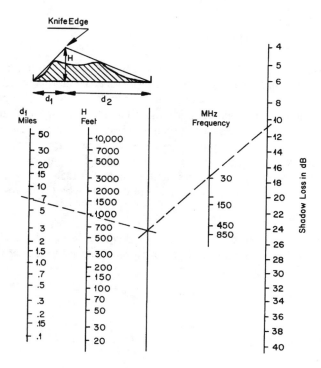

Fig. 15-1 Shadow Loss Relative to Smooth Earth

Shadow Loss of Hills.   Plane-earth propagation assumes a smooth earth. A radio wave passing over a knife edge produces losses due to diffraction. The drawing at the top of figure 15-1 illustrates a knife-edge approximation of a hill. The nomograph in figure 15-1, adapted from Bullington, shows the shadow loss of hills relative to smooth earth. The frequencies represent land mobile radio bands.

---

* Detailed comparisons of various propagation models are given in Section IV of *IEEE Transactions on Vehicular Technology,* Vol. 37, February 1988.

Egli model. The Egli model was published in October 1957 in the *Proceedings of the IRE* by John Egli and gives an overall propagation loss over gently rolling terrain with average hill heights of approximately 50 ft. This model was based on data taken over radials mainly throughout the eastern seaboard and central plains states. The data were measured with receiver-transmitter distances of 40 miles. It is easy to implement and gives an overall indication of propagation. The formula for the Egli model is

$$A_E = 117 + 40 \log D + 20 \log F - 20 \log(H_T H_R)$$

where

A_E	=	Egli model attenuation, dB
D	=	distance, miles
F	=	frequency, MHz
H_T	=	height of transmitter antenna, ft
H_R	=	height of receiver antenna, ft

Okumura model. The Okumura model was published in English in the *Review of the Electrical Communication Laboratory*, September-October 1968. It is based on empirical data compiled into charts that can be applied to land mobile radio system propagation. The basis of this method is the field strength loss relative to free-space in an urban quasi-smooth environment. Okumura defines an urban quasi-smooth environment as an urban area where the undulation height is about 20 m or less with gentle ups and downs. For suburban and open space areas corrections are made from graphs. Additional correction graphs are available for street orientation, local variability, and other factors. The method's details, including all of the many graphs, can be found in the article by Okumura et al. in the References.

Longley-Rice computer model. In 1968 Longley and Rice established a computer program that predicted median values of attenuation referenced to free-space loss. The program requires the following information:

- Frequency
- The transmitting antenna's height
- The receiver antenna's height
- Antenna separation
- Surface refractivity
- Earth's conductivity
- Earth's dielectric constant
- Description of the terrain.

15.3 ESTIMATING COMMUNICATION PROBABILITY

Local area signal variability.   Radio propagation between a base station and a mobile varies with small changes in the mobile location. Propagation models are based on median location data: in a small area 50% have signals below the predicted value. There are also time variations due to seasons, foliage, and so on. However, time variations in a service area are usually minor compared to location variation. Okumura and others have found that the variation of signal levels in a small area have a normal or Gaussian distribution. When the signal level is in decibels, the distribution is called log normal, as figure 15-2 illustrates.

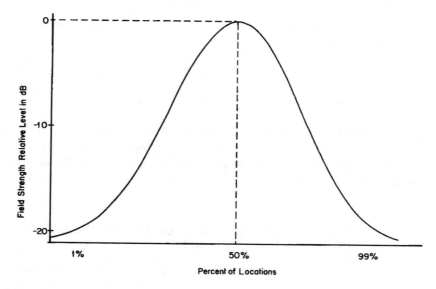

Fig. 15-2 Log Normal Distribution

Standard deviation.   The standard deviation is a measure of the signal's variability in a small area. If enough measurements are taken, the standard deviation can be calculated from the following formula:

$$\text{standard deviation} = \sqrt{\frac{A_1^2 + A_2^2 + A_3^2 + \dots + A_n^2}{n}}$$

where

A_n is the variation from the mean and n is the number of readings.

In land mobile radio systems the standard deviation is given in decibels. The standard deviation values vary with different authors. Table 15-1 shows Okumura's figures:

TABLE 15-1 OKUMURA'S STANDARD DEVIATION VALUES

Suburban and Rolling Hill Terrain

Frequency (MHz)	Standard Deviation (dB)
150	6.75
450	7.5
850	8.25

Urban Area

Frequency (MHz)	Standard Deviation (dB)
150	5.5
450	6.0
850	6.5

Reliability degradation.   The reliability degradation is a loss in decibels that is added to other losses to obtain the desired percentage communication probability. Multiplying the standard deviation by a multiplication factor produces the reliability degradation figure. Table 15-2 shows a list of multiplication factors[*] for various communication probability percentages.

TABLE 15-2 COMMUNICATION PROBABILITY VERSUS STANDARD DEVIATION MULTIPLYING FACTOR

Communication Probability (%)	Standard Deviation Multiplying Factor
90	1.3
95	1.64
99	2.3
99.99	3.73

Communication probability design.   Most land mobile radio systems are designed for 90% probability, whose multiplying factor is 1.3. A communication system operating at 850 MHz has a standard deviation of 8.25 dB according to Okumura. The reliability factor is 8.25 x 1.3, or 10.725 dB. This represents a loss that is added to the propagation path loss to increase the probability percent from 50% to 90%. A designed probability percentage of 95% requires a degradation loss of 8.25 times 1.64, or 13.53 dB. To go

---

[*] The multiplication factor in table 15-2 can be obtained by using a special probability distribution graph that converts the Gaussian bell-shaped graph to a straight line.

from a 90% probability to a 95% probability, an additional 2.81 dB has to be added, which represents almost a twofold increase in power. Similarly, increasing probability from 90% to 99.99% requires a power increase of 20.05 dB. This illustrates the fact that high communication probability percentages will have a very high or impossible cost. Sometimes in public safety communication contracts, specifications call for a 99.99% communication probability. As illustrated above, this is extremely difficult, if not impossible, so a probability of 99.99% should not be used in contracts for public safety communications.

15.4 EXAMINING FACTORS AFFECTING THE RECEIVED SIGNAL

There are many factors that affect the received signal. Among them:

Base transmitter antenna height. Doubling the antenna height produces a gain of 6 dB for plane-earth propagation. This is equivalent to increasing the base transmitter power four times. In general the gain due to increases in base antenna height is 20 log h_2/h_1 where h_1 is the original height and h_2 is the new height. The base transmitter height is the most important variable in a given service area.

Base antenna transmission losses. Table 2-2 gives the typical antenna cable attenuation in decibels per 100 ft for various types of cable and for different frequencies. For example the attenuation for a 7/8-in. foam dielectric cable is 0.5 dB per 100 ft at 160 MHz.

Noise degradation. Man-made noise varies with location and frequency. Low-noise sites are areas isolated from any electrical machinery or high-voltage lines. Moderate noise sites are in suburban areas without industrial electric machinery. High-noise sites are in industrial urban areas. Table 15-3 shows the noise degradation for different environments and frequency bands.

TABLE 15-3 NOISE DEGRADATION IN DB

Environment	VHF Low-Band	VHF High-Band	UHF (450 MHz)	800-MHz Band
Low noise	2	0	0	0
Moderate noise	6	3	1	0
High noise	14	10	4	1

Foliage loss. Table 15-4 shows the loss estimates for dense foliage.

Receiver sensitivity. Service coverage depends on the receiver's sensitivity. Table 15-5 shows the figures used in calculations.

TABLE 15-4 FOLIAGE LOSSES

Frequency Band	Foliage Losses (dB)
VHF low-band	8
VHF high-band	9
UHF (450 MHz)	9.5
800-MHz band	10

TABLE 15-5 RECEIVER SENSITIVITY

Frequency Band	Sensitivity (dB) below 1 W
VHF low-band	−143
VHF high-band	−146
UHF (450 MHz)	−146
800-MHz band	−146

Vehicular antenna gain. The mobile antenna gain depends on the location and type of antenna. A quarter-wave antenna on the center of a vehicle's metal roof has a gain of approximately −1 dB with respect to dipole.

Portable antenna gain. Portable antennas have a much greater loss than vehicular antennas. Table 15-6 gives estimates.

TABLE 15-6 PORTABLE ANTENNA GAIN*

Portable Antenna Type	Gain (dB) above Half-Wave Dipole	
	VHF High-Band	UHF Band
Quarter Wave		
Head Level	−9	−2
Hip Level	—	−14
Helical Coil Spring		
Head Level	−13	−8
Hip Level	−24	−17

* The portable data are adapted from the guide by D. J. Lynch listed in the References.

Building losses for portable operations. The information in table 15-7 is referenced to portables operating at head level in the street outside the building. The readings vary widely with the position in the building. Table 15-7 gives estimates.

TABLE 15-7 BUILDING LOSSES FOR PORTABLE OPERATION

Type of Structure	VHF High-Band	UHF Band
Wood Frame	4	5
Reinforced Concrete and Steel Office Building	30	28
Stucco and Wood	11	15
Shopping Centers	20	20

Signal fading. Terrain such as hills, mountains, and man-made environments (suburban and urban regions) causes long-term fading. Multipath reflections by local scatterers such as houses, trees, and other vehicles causes short-term fading.

15.5 ESTIMATING SERVICE COVERAGE

It is often useful to estimate the coverage for a specific transmitter power, location, and frequency. A modified Egli equation can be used to obtain a quick estimate without a detailed knowledge of the terrain.

Modified Egli method.[*] The modified Egli method is based on 90% probability and uses reliability degradation figures that make the results comparable to other methods. The reliability degradation figures are

- VHF low-band   11 dB
- VHF high-band   14 dB
- UHF (450 MHz) 17 dB
- 850 MHz band   19 dB

The modified Egli formula for estimating the range is

$$d = 10^x$$

where

$$x = 1/40(P_T + G_T + G_R - L_{TT} - L_{RT} - L_P - L_N - 117 - S + 20 \log H_T H_R - 20 \log F)$$

[*] This is adapted in part from "An Approach to Estimating Land Mobile Radio Coverage," by Edward A. Neham, *IEEE Transactions on Vehicular Technology*, Vol. VT-23, November 1974.

d = estimated range, miles
P_T = base transmitter power above 1 W, dB
G_T = base transmitter antenna gain, dB
G_R = mobile transmitter antenna gain, dB
L_{TT} = base antenna transmission cable loss, dB
L_{RT} = mobile antenna transmission cable loss, dB
L_P = reliability degradation loss, dB
L_N = noise degradation, dB
S = mobile receiver sensitivity, EIA SINAD, referenced to 1 W, db
H_T = base station antenna height above average terrain, ft
H_R = mobile antenna height, ft
F = frequency, MHz

Example of the Modified Egli Method. A base transmitter at 155 MHz has an antenna 150 ft above ground level. The tower site elevation above sea level is 100 ft. The average terrain elevation above sea level is 75 ft. H_T, the base station antenna height, is 150 + 100 - 75, or 175 ft. The transmitter's power output, P_T, is 100 W or 20 dB. G_T, the base transmitter gain, is 6 dB and G_R, the mobile antenna, is −1 dB. The base antenna transmission line is 225 ft. L_{TT} is 225 (0.5 dB per 100 ft), or 1.125 dB. L_{RT}, the receiver transmission loss, is 0.5 dB. L_P is 14 dB and L_N for a suburban area is 3 dB. The mobile receiver's sensitivity is −146 dBW. H_R, the receiver mobile height, is 6 ft and F, the base transmitting frequency, is 150 MHz. Substituting these values in the modified Egli formula, $x = 1.30675$ and $d = 20.3$ miles is obtained for the estimated range.

15.6 USING COMPUTER PROGRAMS FOR PROPAGATION PREDICTION

Computer programs predict propagation much more accurately than the modified Egli method. Computer programs are also much quicker than using graphs.

Examining Computer Programs

The computer programs consist of two basic parts. The first is a detailed terrain profile of the service area. The second includes equations for propagation path loss and variables for receiver sensitivity, transmitter output, and so on. A computer plotter can be included in the system to draw the predicted service area's boundaries.

Radio path profile. A topographic database is assembled by dividing the area into small squares or rectangles. The effective terrain height above sea level within each square or rectangle is recorded. A computer accesses the database to construct a radio path profile.

Propagation path loss. A program processes the radio path profile to calculate the propagation path loss. The computations are repeated for all the small squares in the area and the computer produces field strength contours. Entering a particular radio system's variables, such as base station height, power, and so on, enables the computer to predict the service area.

Commercial propagation computer programs. There are commercial program packages available for land mobile radio systems. One type includes a terrain analysis package (TAP). TAP can be used with IBM-PC compatibles to access the National Geophysical Data Center's 30-second point elevation database. The user chooses the rectangular area size and the elevation data to retrieve and plot. A base station's latitude and longitude are entered in the program. Then any desired number of elevation data points are retrieved for a large number of radials. In addition data points can be obtained between any two endpoints, which can be useful for co-channel interference studies. A TAP communications module package interacts with TAP to compute coverage contour and free-space field strength values. Included are obstruction attenuation by the Bullington knife-edge diffraction method.

Another type of commercial propagation computer program integrates the U.S. Geological Survey 1:250,000 scale maps with various propagation equations and data. The programs operate on IBM-PC compatibles. Field strength contours and service coverage can be plotted with these programs. A program using the Okumura method is available with modifications. This differentiates among urban, suburban, and open field areas.

15.7 MEASURING THE SIGNAL QUALITY OF YOUR SYSTEM

There are a great many changing propagation paths as a vehicle moves through the service area. In addition, there are various types of environment such as suburban, urban, open fields, and so on within a service area. All of these situations make precise service coverage difficult and complicated, so measurements are very useful to check the propagation predictions.

Measurement Methods

One method is to record the signal's quality continuously as a vehicle moves through the service area. SINAD is a convenient measure of signal quality that converts numbers into a signal description. For example a SINAD below 12 dB is a poor signal, whereas a SINAD between 12 and 20 dB is a fair signal. Similarly a SINAD between 20 and 30 dB is a good signal. Figure 15-3 shows the method for continuously measuring the SINAD of the service area. The type of mobile receiver that is used in service is installed in a vehicle and connected to a SINAD meter. The SINAD meter output feeds a strip-chart recorder.

There are a number of steps in this method. First, check the communication equipment.

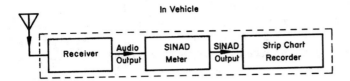

Fig. 15-3 Measuring System Signal Quality

Checking the communication equipment. Measure and record the receiver's 12-dB SINAD sensitivity with a signal generator to make sure that it is within factory specifications. Measure and record the base-station transmitter power under the following modulation conditions: Apply a 1000-Hz tone to either the microphone input or the 600-Ω line input. Adjust the deviation to between 3.0 and 3.3 kHz.

Calibrating the strip-chart recorder. Connect a signal generator to the receiver's antenna input. Figure 15-4 shows the calibration arrangement. The signal generator is tuned to the on-channel frequency with a modulating signal of 1000 Hz and a frequency deviation between 3.0 and 3.3 kHz. Adjust the signal generator's output until the SINAD meter reads 6 dB. Mark this on the strip chart. Similarly mark 12-, 20-, and 30-dB SINAD.

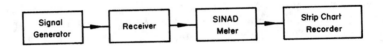

Fig. 15-4 Calibrating the Strip-Chart Recorder

Measuring the signal quality. Adjust the speed of the strip-chart recorder to 15 centimeters per minute (cm/min). Turn on the base transmitter with a modulation tone of 1000 Hz and a deviation between 3.0 and 3.3 kHz. Drive the vehicle at 20 miles per hour—the strip-chart recorder continuously records the SINAD. Figure 15-5 shows a reproduction of a recording, with the signal quality shown for the different SINAD ranges.[*] Each strip is given a number that refers to a list indicating start and stop locations along a street. Notations such as hills, obstacles, and so on are indicated, too.

The first drive should be along the periphery of the desired service area. Then drive through a large number of streets in the service area. Make sure streets that cross the propagation path from the transmitter are included. Areas behind hills and obstructions must be measured. There probably will be a number of narrow dropouts on the chart as the vehicle proceeds.

Analyzing the tests. The strip charts are analyzed for signal quality. Dropout locations are noted. Often moving the vehicle a short distance eliminates these spots.

---

[*] 12-dB SINAD wears on the operator for continuous operations, and systems should not be designed for this rating.

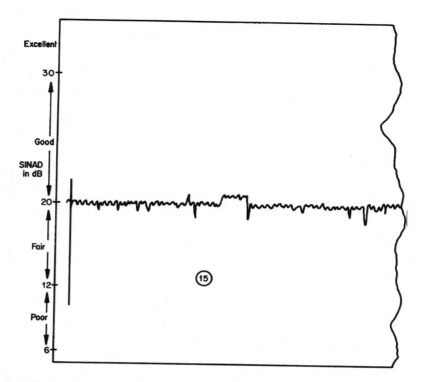

Fig. 15-5 SINAD Chart Readout

15.8 IMPROVING RECEPTION WITH DIGITIZED VOICE

Digitized voice in the near future will begin to replace analog voice in many land mobile radio systems, which will have a marked effect on reception in a service area. With digitized voice the audio quality will be good in areas where the equivalent analog voice is poor. This can be demonstrated easily by removing the antenna from a dual-mode portable near a dual-mode base station. In the analog mode the voice is broken up. In the digitized voice mode the communications are good.

Figure 15-6 shows digitized voice's effect on reception in an idealized service area. This reception improvement will speed the conversion from analog to digitized voice.

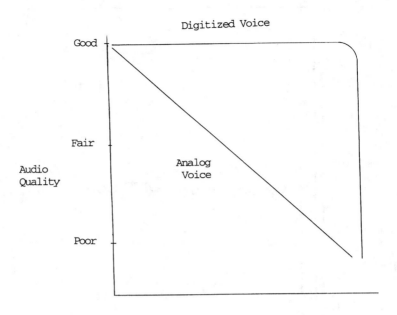

Fig. 15-6
Digitized Voice's Effect on Reception

REFERENCES

Baris, A. P. "Determination of Service Area for VHF/UHF Land Mobile And Broadcast Operations over Irregular Terrain," *IEEE Transactions on Vehicular Technology.* May 1973.

Bullington, Kenneth. "Radio Propagation at Frequencies above 30 Megacycles," *Proceedings of IRE.* October 1947.

_____ . "Radio Propagation for Vehicular Communications," *IEEE Transactions on Vehicular Communications.* November 1977.

Dadson, C. E., J. Durkin, and R. E. Martin. "Computer Prediction of Field Strength in the Planning of Radio Systems," *IEEE Transactions on Vehicular Communications.* February 1975.

Egli, J. "Radio Propagation above 40 MC over Irregular Terrain," *Proceedings of IRE.* October 1957.

Gottschalk, Robert L. "Engineering Consideration for Frequency Coordination." State of Florida, Department of General Services, Division of Communications, Larsen Building, Tallahassee, FL 32301; August 1985.

Hill, C. and B. Olsen, "A Statistical Analysis of Radio System Coverage Acceptance Testing," *IEEE Vehicular Technology Society News*, Feb. 19, 1994.

Longley, A. G., and P. L. Rice. "Prediction of Tropospheric Radio Transmission Loss over Irregular Terrain, a Computer Method." Tropospheric Telecommunications, Boulder, CO 80302; 1968.

Lynch, D. J. "Land Mobile Communications System Design Guide." State of Florida, Department of General Services, Division of Communications, Larsen Building, Tallahassee, FL 32301; July 1978.

Neham, Edward A. "An Approach to Estimating Land Mobile Radio Coverage," *IEEE Transactions on Vehicular Technology*. November 1974.

Okumura, Y., I. Ohmori, T. Kawano, and K. Fukuda. "Field Strength and Its Variability in VHF and UHF Land Mobile Radio Service," *Review of the Electrical Communication Laboratory*. September/October 1968.

Publications without listed authors

"Coverage Prediction for Mobile Radio Systems Operating in the 800/900 MHz Frequency Range," IEEE Vehicular Technology Society Committee on Radio Propagation, published as the entire issue of *IEEE Transactions on Vehicular Technology*, Vol. 37, February 1988.

"VHF and UHF Propagation," *GE Data File Bulletin No. 1003–1*. July 1962.

16

Solving Radio
Interference Problems

Radio interference can seriously degrade land mobile radio systems' operation. This chapter gives solutions to problems in co-channel, adjacent channel, intermodulation, and vehicular ignition radio interference.

16.1 WHAT TO DO WHEN CO-CHANNEL INTERFERENCE PROBLEMS STRIKE

The worst co-channel interference is created when a base station competes with a different mobile radio voting system. Figure 16-1 shows this situation. The base station X is licensed to operate on station Y's mobile frequency. The interference occurs in station Y's fixed voting receivers destroying its communications.

Example of the problem.   In the mid-1970s the Philadelphia Fire Department was licensed to operate a base transmitter on the mobile frequency of the New York City Fire Department in the Brooklyn borough, 153.95 MHz. In figure 16-1 base station X is the Philadelphia Fire Department and system Y is the New York City Fire Department in Brooklyn. In this case Y actually had four voting receivers. Although the distance between Philadelphia and Brooklyn is approximately 70 miles, the interference in Brooklyn was overwhelming. When the Philadelphia station transmitted a routine message every 15 minutes, fire communications were virtually wiped out. The FCC told the departments to work out the problem between themselves.

Solution.   The solution was worked out in five steps.
Step 1: Measuring and analyzing interfering signals at various voting receivers. The interfering signal can be measured by a test voting receiver that has been modified for this

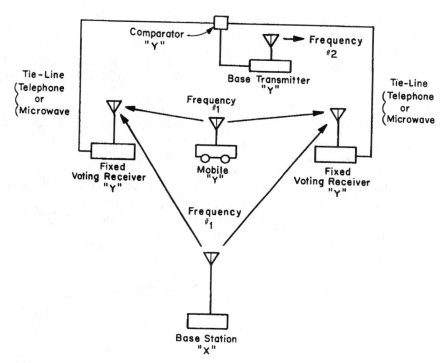

Figure 16.1 Base Station Versus Mobile Voting System

purpose. A meter reading in microamperes is inserted in the first limiter circuit. Care must be taken to avoid saturation. The voting receiver's antenna is connected to the test receiver and readings are taken at all voting receivers. Readings should be taken both during the day and late at night. The largest reading at each site should be used. Table 16-1 shows the results in the Philadelphia-Brooklyn situation.

TABLE 16-1 PHILADELPHIA-BROOKLYN READING RESULTS

Voting Receiver	Interfering Signal μV)
1 (on a very high hill)	10.0
2	0.3
3	0.5
4	3.0

In general, any interfering signal above 1 μV warrants some action at the voting site: either eliminating or using directional antennas or lowering the antenna.

Step 2: Eliminating a voting receiver. The capture ratio for mobile versus base stations is 12 dB (four times the voltage ratio). So voting receiver 1 must receive the Brooklyn mobile with a 40-μV signal to capture the Philadelphia signal. Thus the interfering signal level at site 1 does not lend itself to any corrective procedures. Therefore, receiver site 1 was eliminated. If necessary, two or more voting sites with lower antennas could be substituted to avoid a loss of coverage.

Step 3: Using a directional antenna. Figure 16-2 shows voting receiver 4's relative location. Because of its geographic location, this voting receiver is a candidate for a directional antenna to reduce co-channel interference. Commercial antennas that use a cardioid pattern like the one in figure 16-3 are available. The null, of course, should be pointed in the direction of the interfering signal.

In this case, properly installing the directional antenna lowered the interfering signal from 3.0 to 0.35 μV. The interference was wiped out during the day. However around midnight the Philadelphia signal came back for three or four hours. The next option was to lower the interfering base-station antenna's height.

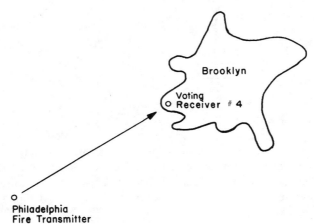

Fig. 16-2 Voting Receiver Candidate for a Directional Antenna

Step 4: Lowering the interfering base-station's antenna. The Philadelphia base-station antenna is located in the crown of the hat of the William Penn statue at the City Hall in Philadelphia. The antenna is 450 ft above sea level. The authorities in Philadelphia cooperated by switching to another base station, where the antenna was 250 ft above ground. Halving the antenna height for distances up to approximately 40 miles reduces the signal by approximately 6 dB. At distances such as 70 miles (as in this case), however, the signal reduction can be much greater. When the Philadelphia antenna height was lowered, the night interference basically was wiped out. However, for one or two days a year in the spring, the interference returned for two or three hours.

Step 5: Adding tone-coded squelch. The final step was to install tone-coded squelch in the New York City Fire Department's radio system. This virtually wiped out all interference from Philadelphia. Note using coded-squelch systems does not stop the interference

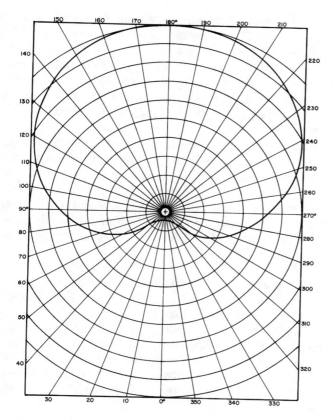

Fig. 16-3 Cardioid Antenna Pattern

if both the interfering base station and the mobile station "victim" are transmitting simultaneously. But when used with the other steps, it does the job.

Summary of steps to reduce co-channel interference, base versus voting systems. The five basic steps to be followed:

1. Measure and analyze the interfering signals at voting receivers.
2. Make a decision to substitute a number of low antenna sites for a high antenna site or use directional antennas for each voting receiver site.
3. Select an appropriate directional antenna. The cardioid is one useful type. There are other types available commercially for special cases.
4. Enlist the cooperation of the interfering base station to lower its base transmitter antenna, if possible.
5. Install a coded-squelch system if it is not already in use. Bear in mind there are only 37 tones in a tone-coded squelch system and there are 2^8 combinations, or 256 tones in an 8-bit digital squelch system. The next co-channel interference problem details the results when two systems have the same CTCSS tone.

Properly Installing a Directional Antenna. Often it is necessary to install a voting receiver directional antenna on a building where there are many other antennas. These can fill in the antenna null and destroy the directional antenna's effectiveness unless certain precautions are followed. First, make sure there are no other antennas or metal in a horizontal plane within a distance of 10 wavelengths (except in the null direction). If the directional antenna is to be mounted with other antennas, one above the other, the separation should be at least 3/4 of a wavelength. If possible, the directional antenna should be mounted below the others.

After the directional antenna has been mounted, someone may install an antenna next to it, completely destroying the pattern. To avoid this look for an offset balcony below the roof. Install the directional antennas on this balcony with the bulk of the building between the interfering signal and the antenna null. The advantages: other people tend to install their antennas on the higher roof instead of the balcony below, so new antenna installations do not interfere with the directional antenna pattern. The bulk of the building helps to shield the antenna from the interfering signal, also. Of course this balcony may not always be available in the needed direction.

Second example of co-channel interference. Sometimes co-channel interference causes havoc without a known source. One day in 1982 the New York City Fire Department suddenly heard its Queens borough base transmitter (154.400 MHz) on the system citywide voting receivers (153.890). This intruding signal's level was great enough to capture vehicular transmissions on the citywide mobile frequency. The next day the New York City Fire Department was advised that the fire department radio station in Manchester, Connecticut, was hearing Queens fire traffic on 153.890 MHz with a tone-coded squelch of 131.8 Hz. Since this was also Manchester's squelch tone, the Queens fire radio communications overpowered the Manchester voting system.

The fact that the Queens base transmitter did not have any tone-coded squelch suggested that an unknown repeater was relaying the Queens signal to Manchester. The receiver of the relay would have to be on 153.890 MHz with a tone-coded squelch of 131.8 Hz. This indicated a newly licensed repeater whose receiver was not using tone-coded squelch until the system was fully operational. A check was made with the local frequency coordinator, who had no record of a repeater being recently licensed with the two frequencies of 153.890 and 154.400 MHz.

After a week the interference on the citywide mobile frequency forced that channel to be abandoned temporarily. Citywide radio traffic was taken up on another frequency. Also, the radio interference at Manchester, Connecticut, became so bad that it installed 10-dB pads in all of its voting receivers, with a resulting sharp loss in sensitivity.

Using a Direction Finder. Because of the situation's seriousness, a vehicle with a direction finder was used to try to locate the culprit. When using a direction finder, it is important to find a high position with a clear view to avoid reflection, which is sometimes difficult in a metropolitan area. Three good bearings are necessary to pinpoint the interfering source's position. In this case one good, strong bearing was obtained, which turned out later to be right on target. However, bearings taken from lower locations were much weaker and subject to reflections, which gave a false position.

Clarifying the Problem. One day the citywide frequency stopped retransmitting Queens fire radio communications. However, the Manchester Fire Department now was receiving interference from an unknown fire department and rescue unit. A call signal was given and the FCC identified the station as one located on a mountain in Hawley, Pennsylvania. The situation was now clarified, as figure 16-4 shows. At first the Hawley repeater had its CTCSS off for its receiver but on for its transmitter. The repeater relayed the Queens fire radio communications to Manchester. After a testing period the relay receiver CTCSS was turned on, and the relaying from Queens to Manchester stopped. The problem now was the interference from a base station in Hawley, Pennsylvania, operating on a mobile frequency to the Manchester Fire Department.

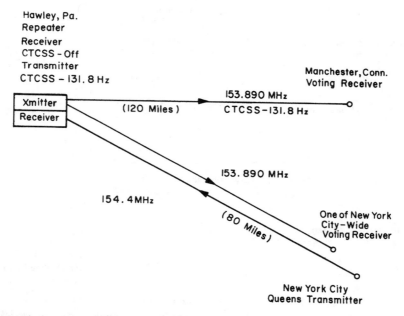

Fig. 16-4 The Solution to the Interference Mystery

Solution. The Hawley Fire Department is a relatively small volunteer unit with only a few vehicles. It interchanged its base and mobile frequencies and the interference ceased. If this had been a system with a large number of vehicles, there most likely would have been strong objections because of the expense involved. In that case you can initiate an action with the FCC. If that fails, the five basic steps discussed previously can be applied.

16.2 REDUCING ADJACENT CHANNEL INTERFERENCE

In large cities voting receivers have to be protected from adjacent and spurious radio interference. There are two methods of protecting voting receivers: a resonant bandpass cavity and RF crystal filters.

Bandpass resonant cavity. Figure 16-5a shows the bandpass cavity resonator. The loops may be rotated to vary the Q, or selectivity. The greater the selectivity, the greater the insertion loss. Figure 16-5b shows the functional equivalent of the bandpass cavity. Varying the inner conductor's height tunes the cavity to the exact frequency. A resonant bandpass cavity at 100 kHz from the cavity's center frequency can have an attenuation of 12 dB with an insertion loss of 3dB. Refer to manufacturers' specifications for actual figures.

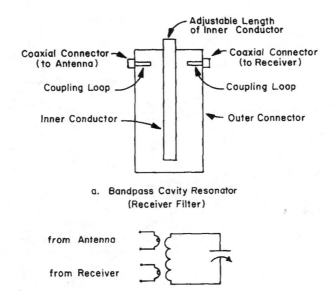

a. Bandpass Cavity Resonator
(Receiver Filter)

Fig. 16-5 Bandpass Cavity Filter for a Voting Receiver

RF crystal filters. In the VHF band a crystal filter can be used between the antenna and the receiver to reduce adjacent channel signals. To compare RF crystal filters with the bandpass cavity: at 100 kHz from its center frequency the crystal filter in VHF high-band can result in an attenuation of 40 dB with an insertion loss of 6 dB. Again refer to manufacturers' specifications for actual attenuation and insertion loss. Crystal filters are not available above the VHF high-band.

16.3 KEEPING INTERMOD AWAY FROM YOUR DOOR

To be able to reduce intermodulation interference, it is necessary to understand the different types because each has its own treatment.

What Is Intermodulation Interference?

Intermodulation is the process by which two or more radio signals impressed on a nonlinear device produce additional frequencies. There are three general types of intermodulation: transmitter, receiver, and, less frequently, external devices such as corrosion on antenna guy lines.

Intermodulation products are classified by their "order" (second, third, fourth, fifth, etc.) and the frequencies that give rise to the product are designated A, B, C, etc. The order of any product is equal to the sum of the terms. For example, $(A - B)$ and $(A + B)$ are second-order products. $(2A - B)$ and $(2B - A)$ and $(2B + A)$ and $(2A + B)$ are all third-order products. This may be extended through third-, fourth-, fifth-, and so on, order products. Practically speaking, the worst offenders are the third-order difference products $(2A - B)$ and $(2B - A)$. Transmitter and receiver intermod have different causes and therefore must be treated differently.

Transmitter intermod interference. Figure 16-6 illustrates transmitter intermod interference. The radiated signal from transmitter B enters the antenna of nearby transmitter A and proceeds down the transmission line to the final stage of the transmitter. Here it combines with the second harmonic of transmitter A to form C, intermod frequency $(2A - B)$. The intermod is radiated from transmitter A. Note that the intermod $(2B - A)$ is not radiated from transmitter A since there is an FCC-required second harmonic filter that attenuates the radiated frequency $(2B)$ by 80 dB. This is a useful way to determine which transmitter is radiating the intermod. An example of transmitter intermod that occurred in the New York City Fire Department illustrates this.

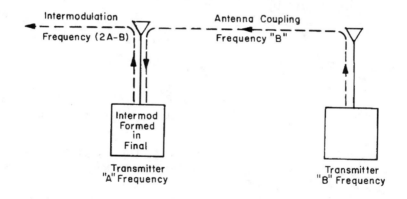

Transmitter Intermodulation

Fig. 16-6 Transmitter Intermodulation*

* While transmitter "A" radiates the intermodulation frequency $(2A - B)$, transmitter "B" radiates $(2B - A)$. The radiated intermodulation frequency $(2B - A)$ has been omitted for simplicity.

Sec. 16.3 Keeping Intermod away from Your Door

Example of Transmitter Intermod. The New York City Fire Department received a report that there was radio interference on 154.070 MHz. The interference was observed directly on the spectrum analyzer only when two of the department's transmitters were on simultaneously. The transmitter on 154.250 MHz was designated as A and the transmitter on 154.430 MHz was B. The two third-order possibilities then were examined.

$$2B - A = 2(154.430) - 154.250 = 154.610 \text{ (not the interference signal)}$$

and

$$2A - B = 2(154.250) - 154.430 = 154.070$$

Thus it is the last equation that gives the interfering signal. Also, it shows that transmitter A on 154.250 is radiating the intermod. The transmitter frequency multiplied by 2 is the one radiating the intermod for the reason given previously. This means this transmitter should be modified to reduce the reported interference on 154.070.

Cure for Transmitter Intermod. The intermod can be reduced by an isolator, the three-port unidirectional device shown in figure 16-7. Radio frequency can go only in a clockwise direction from port 1 to port 2 and from port 2 to port 3. Thus the desired energy from the transmitter at port 1 goes to port 2 and up to the antenna, while the undesired energy from an extraneous transmitter at port 2 can go only in a clockwise direction to the dummy load, where heat absorbs it. An isolator without a dummy load is called a circulator.

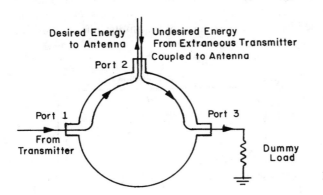

Fig. 16-7 Isolator

One isolator attenuates the undesired signal by 25 dB with an insertion loss of 0.5 dB. Two isolators in tandem give an attenuation of the intermodulation power by 50 dB with an insertion loss of 1.0 dB.

An isolator is a nonlinear device that generates harmonics. There are two methods of reducing harmonics isolators produce. One is a low-pass filter inserted between the antenna transmission line and the isolator. This method works fine when using a duplexer. However, there can be a problem when a relay is used to switch the antenna from transmitter to receiver. In this case the capacitor in the filter may be punctured by a switching transient putting the transmitter off the air. To avoid this use an isolator with a quarter-wave (electrical) shorted coaxial stub across the isolator, where it is connected to the antenna.

Figure 16-8 shows two isolators connected together with a low-pass filter. A quarter-wave shorted coaxial stub can be used in the case of an antenna switching relay.

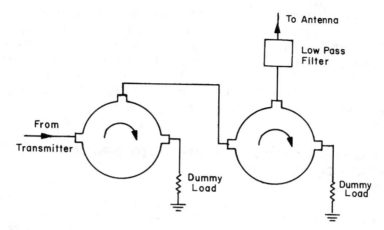

Fig. 16-8 Two Isolators in Tandem

Receiver Intermod. In receiver intermodulation two or more separate transmitted signals enter the receiver and the intermodulation products are formed in the receiver RF amplifier or mixer, as figure 16-9 shows.

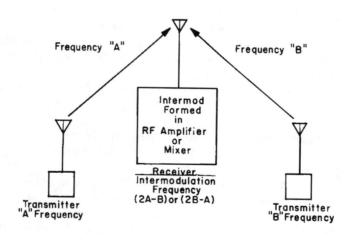

Receiver Intermodulation

Fig. 16-9 Receiver Intermodulation

Cure for Receiver Intermod. The methods of reducing receiver intermod are the same as those for adjacent channel interference: namely using bandpass cavity resonators or RF crystal filters, if necessary. The bandpass cavity resonator or crystal filter is connected between the receiver and antenna.

Determine the type of intermod. Since the remedies for receiver intermod are different from the transmitter intermodulation remedies, it is important to have a test to distinguish between these two types of intermodulation. For third-order receiver intermodulation, a 10-dB RF attenuator at the receiver input reduces the output by 30 dB. If the intermod is the transmitter type, a 10-dB attenuator at the receiver input results in a 10-dB reduction of the output. With this test, the proper remedy then can be applied. Another possible source of interference is ignition vehicle noise.

16.4 CONTROLLING IGNITION AND OTHER INTERFERENCE SOURCES IN THE VEHICLE*

Ignition and other noise sources can degrade mobile reception in low signal-to-noise areas even though the interference is not apparent. It is therefore important to understand the mechanisms that produce radio interference from ignition noise and other vehicular sources to understand how to deal with them.

Examining Ignition Interference Sources

In nonelectronic ignition systems there are three separate sources of radio interference that appear to be one because they occur within milliseconds of each other. First, the air gap between the distributor rotor and stator breaks down, extending the high voltage to the spark plug. Second, the spark plug fires, and third, the distributor breaker points close to pass current through the high-voltage coil. The last step is the least significant source of radio interference. In electronic ignition systems the solid-state switching replaces the distributor breaker points and condenser of the nonelectronic system. However, the total interference is not reduced significantly.

Suppressing electromagnetic interference from ignition systems. SAE Standard J551g, "Limits and Methods of Measurements of Radio Interference Characteristics of Vehicles and Devices (200–1000 MHz)" is the voluntary suppression standard the auto industry uses. To meet these limits auto manufacturers usually incorporate resistor spark plugs and special cables. In the resistor spark plug an internal resistor built into the spark plug body attenuates the electromagnetic interference (EMI). Special cables some auto manufacturers use are high-voltage cables with a distributed resistance from 3000 to 7000 Ω/ft. Some use a magnetic cable consisting of a magnetic material surrounded by a coil of wire. All of these methods may reduce the interference until it has no effect in a high signal-to-noise area. However, in a marginal area it may degrade receiver perfor-

---

* Much of the material in this section has been adapted, with modifications, from the *NBS Special Publication 480-44* by H. E. Taggart, listed in the References.

mance. Some special techniques to further reduce automobile ignition noise are described in detail in an article by R. H. Shepherd et al. listed in the References. The article includes details on building low-pass filters into standard resistor plugs and the distributor.

Other Sources of Automotive Interference

The alternator, electric motors, warning buzzers, and turn indicators can cause problems also. RF energy from radio transmissions can affect computers that are used to control functions in the vehicle. All of these interfering sources have two basic paths: radiating and conducting.

Controlling nonignition interference.
A vehicle made of metal may have sections that are not electrically connected to the rest of the vehicle. These electrically isolated parts may radiate radio energy if they are approximately one quarter-wave long. For example, paint may isolate the hood from the rest of the car so the hood may act as an antenna at certain frequencies. This also may be true of bumpers, tailpipes, and so on. You can control the radiation path and increase a metal car's shielding by using grounding straps to connect one metal section to another. Using line filters and/or bypass capacitors suppresses EMI in a vehicle, also. The alternator whine heard in a vehicle radio is an example. Eliminate this interference by using an L-filter in the positive lead to the radio transceiver. The filter is made of an RF choke and a capacitor that is connected to a good vehicle ground. The choke is wound with heavy-gauge wire to pass the current the transceiver requires. The capacitor should have the proper voltage rating for the vehicle's electrical system. Alternator filters are commercially available.

The interference from various motors and buzzers in the vehicle can be handled by using bypass capacitors. For example, suppress the interference from a windshield wiper motor by using a coaxial capacitor in the power leads to the motor. A capacitor between the motor brushes and the vehicle ground suppresses the EMI at the source. The radio transmitter itself may interfere with computerized vehicle systems. A true-life example: The RF from the vehicle transmitter has activated the buzzer and warning lights for temperature gauges. The RF energy gets into the vehicle's transistorized sensor board. Two bypass capacitors on the sensor board eliminate the problem.[*]

Testing for vehicle receiver degradation.
A vehicle receiver operating in a good signal-to-noise area may not experience any noticeable degradation due to EMI in the vehicle. However, in a marginal area a receiver may experience enough degradation to change a barely acceptable signal to a poor signal. The following test measures this degradation:

The vehicle should be placed in an EMI quiet area away from high-voltage lines, electrical machinery, and radio signals. Figure 16-10 shows the equipment arrangement in the vehicle. A signal generator with a modulating signal of 1 kHz at a deviation of 3.0 to 3.3 kHz is tuned to the receiver channel. The signal generator is coupled loosely (20 to 30

---

[*] Details on microprocessor EMI solutions can be found in the article by D. White et al. in the References.

dB) to both the antenna and receiver. The antenna is coupled directly to the receiver through the coupling box. A SINAD meter is connected across the loudspeaker.

Before the actual test, check the coupling of the signal generator to the receiver by the following procedure: Turn off all electrical equipment except the test equipment. Feed the signal generator directly into the receiver and record the signal generator voltage for 12 dB SINAD. Then connect the signal generator, antenna, and receiver to the coupling box. Increase the signal generator output until the SINAD meter again reads 12 dB and record the signal generator output. This output should be between 10 and 30 times the signal generator output without the coupling box. The coupling can be adjusted and the actual degradation test then can be carried out.

First, turn off all electrical equipment in the vehicle except the test equipment. Connect the antenna, signal generator, and the receiver to the coupling box. Adjust the signal generator until the SINAD meter reads 12 dB. Turn on the ignition and record the SINAD meter reading. If the SINAD meter reading is still 12 dB, there is no degradation due to the ignition system. A 12 dB SINAD reading represents the lowest acceptable useful signal. A decrease from the 12-dB SINAD means that a marginally useful signal has been reduced to a poor signal. Turn on all the electrical devices in the vehicle one by one to determine which devices cause a degradation of signal. When a specific piece of equipment is identified, the EMI suppressing method can be used.

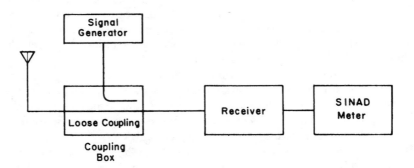

Fig. 16-10 Receiver Degradation Test

REFERENCES

Cohn, Jona, and Rick Chander. "System Improvements Through the Use of RF Crystal Filters," *IEEE Transactions on Vehicular Communications.* March 1966.

Guthrie, A. K. "Living in a Crowded World." Seminar Notes, GE Mobile Radio Technical Training, ECP–877A, 1982.

Kuo, Wey-Chaung. "Suppression of Radio Frequency Interference at the Distributor Rotor Gap," *IEEE Transactions on Vehicular Technology.* May 1979.

Shepherd, R. H., J. C. Gaddie, and D. L. Nielson. "New Techniques for Suppression of Automobile Ignition Noise," *IEEE Transactions on Vehicular Technology.* February 1976.

Taggart, Harold E. "Methods of Suppressing Automotive Interference," NBS Special Publication 48–44. November 1981. A single copy can be obtained from National Bureau of Standards, Building 221, Room B157, Gaithersburg, MD 20899.

White, D., K. Atkinson, and J. Osburn. "Taming EMI in Microprocessor Systems," *Spectrum*. December 1985.

Publications without listed authors

FCC Report of the Advisory Committee for the Land Mobile Radio Service. Vol. 2. 1967.

"Intermodulation Interference in Radio Systems," *Bell System Technical Reference PUB 43302*. November 1972. This can be obtained from AT&T Customer Information Center, Commercial Sales Representative, P.O. Box 19901, Indianapolis, IN 46219.

"Limits and Methods of Measurement of Radio Interference Characteristics of Vehicles and Devices (20–1000 MHz)," SAE Standard J551g. Society of Automotive Engineers, Inc., Warrendale, PA, May 1979.

17

Communications in Enclosed Areas

This chapter covers the limitations of conventional radio equipment in enclosed areas and methods of improving communications. Included are leaky cable design, using crossband vehicular repeaters in high-rise building fires, LAN radio systems within enclosed areas, and personal communication networks (PCN) linking radio systems within a building to the outside.

17.1 UNDERSTANDING WALKIE-TALKIE LIMITS IN ENCLOSED AREAS

Ordinary radio communication equipment that uses walkie-talkies does not work well in enclosed spaces such as tunnels, subways, and large buildings.

Limitations in tunnel communications. If a tunnel is treated as a waveguide, there is a certain frequency, related to the cross-section dimensions, called the cutoff frequency. Below this cutoff frequency there is no waveguide transmission. The cutoff frequency may be in the microwave region. In practice radio communication using 1-W walkie-talkies at 150 MHz in a subway tunnel in New York City is limited on the average to approximately 200 ft. Where the subway tunnel turns or dips, the communication may be even more limited. As the frequency goes up, the distance of communications in a straight line increases. Ordinary radio communications in a tunnel must depend on special equipment such as leaky cables strung throughout the length of the tunnel that act as a long antenna. For emergency services such as fire departments, there are other systems that may be used when leaky cable is not available.

Limitations in high-rise buildings (above 10 floors). Communicating by walkie-talkie radio for security and other services in high-rise buildings is difficult if not impossible, especially between the lower and highest floors. Communications are mostly vertical in a high-rise building. For a walkie-talkie antenna held vertical, little radio energy goes straight up so systems such as leaky cable run in a vertical shaft must aid vertical communication. Here again, where there is no available leaky cable, special methods may be used for emergency services such as fire departments.

Limitations in shopping malls. In a large shopping mall communications for inside and outside security forces are difficult. In some cases a repeater with an outside antenna solves this problem.

17.2 USING LEAKY CABLE SYSTEMS

Leaky cables are used in subways, tunnels, and high-rise buildings to enable walkie-talkies to be used in security and fire communications.

Leaky Cable

The Andrew Corp.'s RADIAX leaky cable is used often. Figure 17-1 shows a cutaway section of the cable. There are a series of slots in the outer conductor of the coaxial cable that allow radio waves to enter and to leave the cable.

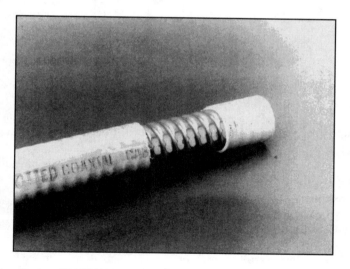

Fig. 17-1 Cutaway Section of RADIAX Leaky Cable*

* Courtesy of Andrew Corp.

High-rise building installation. Figure 17-2 shows a high-rise building installation. Communication between two walkie-talkies using a leaky cable is not feasible without the repeater because of losses in the cable.

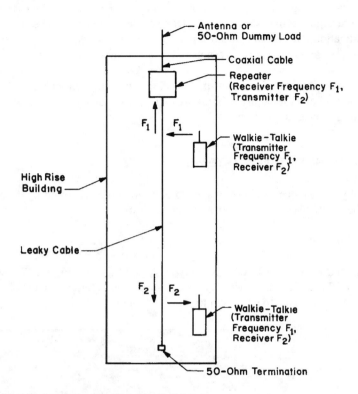

Fig. 17-2 Leaky Cable Installation in a High-Rise Building

A console for a security headquarters can be connected to the repeater with a telephone and control line. This enables supervisory personnel to talk to personnel on different floors.

The leaky cable should be installed vertically in a nonmetallic elevator or utility shaft. To provide uniform coverage the shaft should be as close to the center of the building as possible.

If communications are desired outside the building also, an antenna is installed on the roof. Otherwise, a dummy load is used to keep electromagnetic waves within the building.

Subway installation. Figure 17-3 shows an example of a leaky cable installation in a subway system. The cable itself is installed on the subway tunnel wall. Each subway station has its own independent installation. Figure 17-3 shows two systems—the train dispatching system and a transit police communication system—each on its own frequencies. A combiner joins the two. A splitter then feeds two opposite legs of the leaky cables.

Each leg ends in a 50-Ω termination. A subway train has a transceiver with an antenna. The subway train maintains communication with the train dispatching headquarters by transmitting and receiving through the leaky cable. Telephone lines from the station transceiver, T, connect to a console in headquarters. The console operator at the headquarters knows automatically which subway station is communicating. Transit police using duplex walkie-talkies can communicate with each other through repeater R, and they can also communicate with their headquarters.

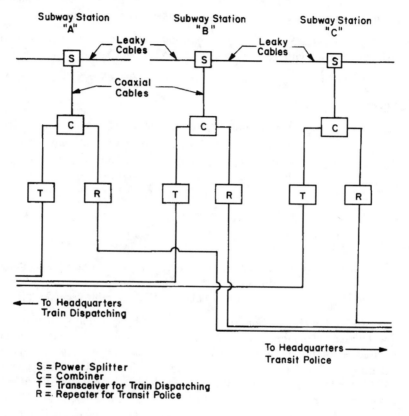

S = Power Splitter
C = Combiner
T = Transceiver for Train Dispatching
R = Repeater for Transit Police

Fig. 17-3 Example of Leaky Cable Subway Installation

Each time another service, such as the fire department, emergency medical services, and so on, is added to the system, the losses add up. The design of the overall system must account for the losses.

A leaky cable system design.　　There are two types of leaky cable losses: coupling loss and attenuation loss per unit length of the cable.

Coupling Loss. Coupling loss is the average difference in decibels between signal level in the cable and the power an antenna 20 ft from the cable receives. A tolerance of ±10 dB is used.

Attenuation Loss. Attenuation is specified in decibels per 100 ft or 100 m. Attenuation goes up with frequency and down with increases in the cable's diameter. One-half- and 7/8-in.-diameter cables are in use.

Other System Losses. Other losses include coaxial cable attenuation, power splitter losses, and other component losses, such as those from combiners.

Leaky Cable System Design Equation. The general design equation for walkie-talkies operating with leaky cables is

$$T_X - A_L - L_C - L_A - PS_L - COMB_L - COAX_L - SU = R_P$$

where

T_X	=	walkie-talkie transmitter output referenced to 1 W, dB
A_L	=	portable antenna loss referenced to a half-wave dipole, dB
L_C	=	leaky cable coupling losses measured at 20 ft with a half-wave dipole, dB
L_A	=	leaky cable attenuation, dB
PS_L	=	power splitter losses, dB
$COMB_L$	=	combiner losses, dB
$COAX_L$	=	coaxial cable losses, dB
SU	=	system use factor (this takes care of extra losses in system operation; -20dB is recommended)
R_P	=	received power below 1 W at the cable receiver, dB

Using the Design Equation. A subway leaky cable system in the 150-MHz frequency range is planned using a 7/8-in. RADIAX cable. The longest single run of cable to be used is 2000 ft. The design includes two separate operating systems: one for train dispatching and one for transit police operations using walkie-talkies. The worst situation is when a walkie-talkie is transmitting at the end of a run of 2000 ft. of cable. The walkie-talkie has a transmitter output of 1 W, or T_X is 0 dB referenced to 1 W. The walkie-talkie antenna is the helical coil spring type held at hip level. The portable antenna loss, A_L, is -24 dB from table 15-6. The leaky cable coupling loss, L_C, is –70 dB, given by the manufacturer. The cable manufacturer lists the leaky cable attenuation as 0.6 dB per 100 ft. For 2000 ft, L_A, the total cable attenuation is –12 dB. The power splitter loss, PS_L, is –3 dB. The combiner loss, $COMB_L$, for two systems is also –3 dB. The coaxial cable loss, $COAX_L$, is –2 dB.

T_X	=	0 dBW
A_L	=	−24 dB
L_C	=	−70 dB

$$L_A \quad = \ -12 \text{ dB}$$
$$PS_L \quad = \ -3 \text{ dB}$$
$$COMB_L \ = \ -3 \text{ dB}$$
$$COAX_L \ = \ -2 \text{ dB}$$
$$SU \quad \ = \ -20 \text{ dB}$$
$$R_P \quad \ = \ -134 \text{ dBW}$$

The receiver at the other end of the cable has a sensitivity of −143 dBW. The system performance margin is the difference between the received power and the receiver sensitivity, or 9 dB.

Daisy chains.   The daisy chain is another method that uses leaky cables in subways and tunnels. One disadvantage of the conventional leaky cable installation shown in figure 17-3 is that a large number of transceivers must be used. Because of the leaky cable attenuation, the cable must be broken up into sections, each with its own set of transceivers. In a subway system, the number of transceivers can be very large. The daisy chain, shown in figure 17-4, minimizes the number of transceivers. Amplifiers compensate for the cable attenuation in daisy chains, which are used mostly in Europe.*

The daisy chain in general eliminates a great deal of equipment such as transceivers and power splitters. However, it requires a large number of amplifiers and has a greater chance of catastrophic failure. The single daisy chain in figure 17-4a has one obvious disadvantage. One break in the cable or a bad amplifier destroys the system's operation. The double daisy chain in figure 17-4b minimizes this by using two systems.

17.3 SUBWAY EMERGENCY COMMUNICATIONS WHEN LEAKY CABLES ARE NOT AVAILABLE

Sometimes leaky cables are not available in subways for emergency operations such as fire and EMS. In these cases special systems have been developed to provide emergency communications. A portable sound power telephone system and a portable repeater are two of these special emergency systems.

Portable sound-power telephone systems.   One portable sound-power telephone system consists of a 30-lb portable reel. Each reel contains 500 ft of a two-conductor cable. The reel is carried into the emergency site and the wire is unrolled automatically as the reel is carried in. One reel can be connected to another for a total distance of more than five miles. Sound-powered headsets are connected at each end of the sound-powered cable. No external source of power is needed and a loud voice speaking directly into the headset's sound tube generates about 0.1 V. It is possible for the manufacturer to modify one of the headsets for throat operation so it can be used with breathing apparatus

---

* A study of underground radio communications in Europe was made by R. A. Isberg, listed in the References.

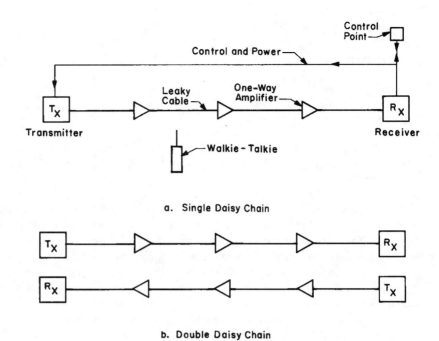

a. Single Daisy Chain

b. Double Daisy Chain

Fig. 17-4 Daisy Chain Formations

in smoke situations. The headset modification lowers the audio signal, but it is still usable. The sound-powered system's advantage is the ability to talk from the street command post down into a tunnel without external power. The disadvantage: There is a long trailing wire that can get in the way.

One method of avoiding this disadvantage is to install a sound-powered cable with two AWG No. 16 conductors in the subway. The cable runs from a junction box in the sidewalk next to an exit down to the subway tunnel. Junction boxes are installed every 200 ft in the tunnel. A firefighter at the fire scene in the subway plugs a sound-powered headset into the nearest junction box. The fire chief in the street above plugs a headset into the junction box and communicates with the firefighter. This system has been installed on a trial basis in one subway station in the New York City subway system. This type of installation is much cheaper than installing a separate leaky cable system for the fire department.

Portable repeater.   A portable repeater can be carried into the subway tunnel to extend the range of walkie-talkie operation.

17.4 USING CROSS-BAND VEHICULAR REPEATERS IN HIGH-RISE BUILDING FIRES

The New York City Fire Department developed the cross-band vehicular repeater system in 1979 for communications in high-rise building fires, where leaky cables are not available. In New York City, the fire chief in charge of the high-rise fire goes to a lobby command post adjacent to a fire command panel on the first floor. Among other things the fire command panel shows the floor where the fire alarm has been activated.

As mentioned previously, communications straight up or down from a vertical antenna are extremely limited. To avoid this problem a battalion chief's car is equipped with a cross-band repeater. Figure 17-5 illustrates the cross-band repeater's basic operation when used in high-rise audio communication. The chief in charge, located at the lobby command post, transmits to his car on a special 1-W UHF walkie-talkie. A special UHF transceiver receives the message, which is passed automatically to the regular two-way VHF radio, also located in the car. The VHF radio is equipped with an additional channel, which is a citywide frequency for high-rise operations, operating at a frequency of 154.430 MHz. The transmission then is boosted to 45 W and, in turn, transmitted to the top floors of the building. When this system is used, the transmission does not have to penetrate all the steel and concrete of the different floors of the building.

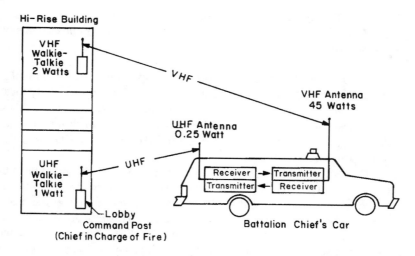

Fig. 17-5 Cross-Band Repeater for Fire Radio Communications in a High-Rise Building

Installing the cross-band repeater. Install the UHF transceiver in the battalion vehicle on top of the regular VHF transceiver. Mount the two antennas shown in figure 17-5 at least 2 ft apart on the battalion vehicle's roof. Install a battery-charging unit for the 1W UHF walkie-talkie in the front of the battalion vehicle, next to the seat. When the UHF walkie-talkie is in this charger, its battery is being charged and the cross-band repeater is inactivated.

The chief turns the regular two-way radio in the car to the high-rise channel to operate the cross-band repeater. The chief then takes the UHF walkie-talkie out of the charging unit in the car, which automatically activates the repeater and inactivates the repeaters in any nearby vehicles. This feature is incorporated so that two or more repeaters do not interfere with each other. (The equipment has a built-in system that gives priority to the last vehicular repeater in operation.) The chief then proceeds to the lobby of the high-rise building.

Positioning the vehicular repeater. The battalion vehicle should be placed a minimum of 60 ft from the high-rise building. This usually means parking the car across the street or down the street from the building.

Using a fixed repeater. In a congested area with many high-rise buildings it may be more feasible to install the cross-band repeater in a tall building. This is particularly true where it is difficult to park the cross-band vehicular repeater in the most advantageous position. The fixed cross-band repeater operation is similar to the vehicular repeater.

17.5 NEW DEVELOPMENTS IN ENCLOSED AREA COMMUNICATIONS

There are two additional developments in enclosed areas: using radio systems in local area networks (LAN) within a building and using PCN to link radio systems within a building to the outside.

LAN radio systems within a building. In a LAN system information usually is passed to different desktop computers by wire. Changes of activity location makes permanent wiring for LAN impractical in some situations. Radio systems that satisfy FCC Part 15 Unlicensed Rules have been developed. These are 1 watt systems operating in 902–928, 1850–1900, 2,400–2,483.5 and 5,725–5,850 MHz. They use CDMA with direct sequence spread-spectrum.

One of the problems is multipath propagation within the building, which using a rake receiver can solve. The rake receiver consists of eight receivers in parallel all time-delayed from each other. The output is fed into a correlator that compares the receivers and selects a signal without echoes from bounces off walls and objects. All eight of the rake receiver's components are built into the CMOS chip. The IEEE is working on Standard 802.11 for wireless LANS and the Europeans are preparing a Hyper-LAN specification.

Personal communications network. In PCNs a cellular network with tiny cells connects radio communications within a building to the outside public telephone systems, as figure 17.6 shows. PCN can use satellite communications, paging, and cellular systems. PCN is being developed in the United States and in the European community. In Europe it is known as universal mobile personal telecommunications services (UMPTS). UMPTS encompasses indoor wireless, cellular, telepoint, and intelligent networks.

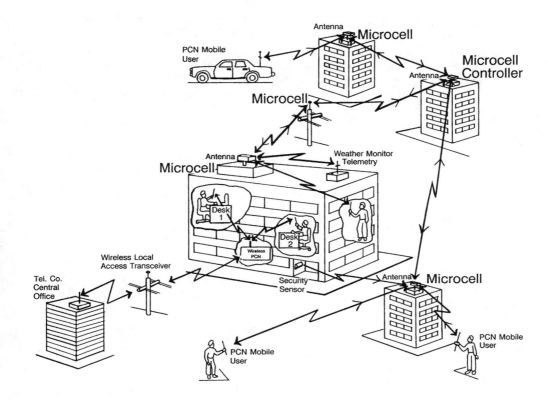

Fig. 17-6 A PCN System[*]

REFERENCES

Brennan, G. R. "Slotted Coax Solves Tunnel Radio Problem," *Communication News*. June 1975.

Isberg, R. A. "A Study of the Technology of Underground Radio Communication in Europe." Presented at American Public Transit Association 1980 Radio Transit Conference. R. A. Isberg, P. E., consulting communications engineer, 1215 Henry Street, Berkeley, CA 94709.

Jakes, William C., Jr. *Microwave Radio Communications*. New York: John Wiley & Sons, Inc., 1974, pp. 110–112.

Kobb, Bennett Z. "Personal Wireless," *Spectrum*. June 1993.

Seidel, Scott and Ted Rappaport. "Path Loss Prediction in Multi-Floored Buildings at 914 MHz," *The Journal, Electronics Letters*. July 18, 1991. Part of Scott Seidel's Ph.D. dissertation at the Mobile & Portable Radio Research Group of Virginia Polytechnic Institute and State University.

_____ . "914 MHz Path Loss Prediction Models for Indoor Wireless Communications in Multi-Floored Buildings," *IEEE Transactions on Antennas and Propagation*. February, 1992.

[*] By permission of InterDigital Communication Corporation.

Takai, Hitoshi. "In-Room Transmission BER Performance of Anti-Multipath Modulation PSK-VP," *IEEE Transactions on Vehicular Technology.* Vol. 42, No. 2. May 1993.

Publications without listed author

"Radiax™ Slotted Coaxial Cable System Design Considerations," *Andrew Bulletin 1058A.* Andrew Corporation, 10500 West 153rd Street, Orlando Park, IL 60462; 1979.

18

Automatically Locating Your Vehicles

Vehicles in land mobile radio systems are constantly moving and it is important to know the vehicles' positions. Locating vehicles by radio messages consumes time and is not the best way to manage personnel resources. AVL gives the continuous location of a vehicle, which is especially useful when used with CAD. AVL can aid in quickly matching addresses of incidents in public safety radio services with available units. A visual display with AVL allows the dispatcher to see precisely where each vehicle is at all times.

This chapter discusses the general methods of locating a vehicle, including sign posts and hyperbolic and circular trilateration. Specific AVL methods in land mobile radio systems are described, including dead reckoning, Loran C, and satellites.

18.1 UNDERSTANDING THE BASIC TYPES OF AVL

Sign posts. With sign posts the vehicle uses a low-powered transmitter to send a vehicle identity code to a fixed receiver located along the street in known fixed positions. The vehicle identification number then is sent by wire to headquarters. This method can use police call boxes or fire alarm boxes in the public safety radio services.

Another method is to place small microwave transmitters every few blocks. The microwave transmission gives the location of the microwave transmitter to the vehicle, which then transmits the information by radio to the dispatching center. Sign posts are especially useful for vehicles such as buses that follow fixed routes.

Dead reckoning. In dead reckoning, instruments in a vehicle keep track of the direction and distance traveled from a known starting point. A computer calculates the vehicle's position on a continuous basis.

Trilateration radio location.   Trilateration is based on the fact that radio signals travel with a constant velocity. Knowing the time a radio signal takes to propagate over a given path allows you to calculate the length of the path. Many of the AVL and navigation systems use trilateration. There are two types of modulation in AVL trilateration: pulse and phase, or sine-wave, modulation. Pulse modulation has an advantage over phase modulation in that multipath propagation does not affect pulse modulation as much. Pulse trilateration is more accurate than phase trilateration. Pulse trilateration's disadvantage is that it requires more bandwidth than phase trilateration.

In general two types of communication channels are used with trilateration systems: a data channel and a clock synchronization channel. The minimum number of fixed sites for vehicle location systems that operate by measuring propagation is three. However, additional stations may be used to increase position accuracy, and the system then is called multilateration instead of trilateration. There are two general types of trilateration systems in use: circular and hyperbolic.

Circular Trilateration Systems.   Circular trilateration systems require three or more fixed receivers. All three receivers and the transmitter in the vehicle must have access to synchronized clocks. Then by noting the time of transmission from a transmitter and the time of its reception in the receiver, the distance can be readily computed from the known velocity of electromagnetic waves. The vehicle then lies on a circle with a fixed receiver at the center and a radius equal to the computed distance. Two circles intersect at two places, so it is necessary to have three circles—one for each of the three receivers—to obtain the mobile position. This is illustrated in figure 18-1. Three fixed transmitters with a receiver in the vehicle can replace the three fixed receivers.

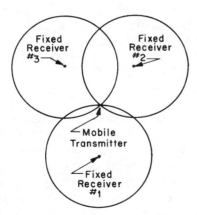

Fig. 18-1 Circular Trilateration System

Hyperbolic Trilateration Systems.   In hyperbolic trilateration systems a master station and a slave station transmit a fixed time-difference apart. A series of hyperbolic curves is generated on a map. Each curve represents a specific time difference between the master and the slave station. A mobile receiver determines the hyperbolic curve it is on by measuring the time difference between a pair of transmitters. Using two pairs of transmitters, it can get an intersection point, as figure 18-2 shows. A third pair of transmitters gives a more precise location. There are a number of hyperbolic systems in use: Loran C, Omega, and Decca.

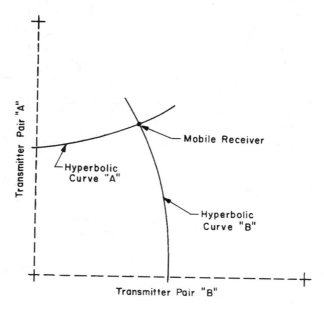

Fig. 18-2 Hyperbolic Trilateration System

Loran C. Loran (long-range navigation) is a pulse trilateration government-operated system transmitting at 100 kHz. A chain consists of a master transmitter and two or three slave transmitters. The transmitters operate at 0.5 MW in the 100 kHz region. At the present time Loran C transmitters cover the 48 contiguous states. The Loran C stations' locations are shown in figure 18-3. Loran C was originally designed for ship and aircraft navigation. However, recently it has been used for automobile location.

Omega. The Omega system is used for ship navigation. It operates in the 10- to 14-kHz very low frequency band and is a phase system.

Decca. This is a European system developed in England. It is a phase system similar to Omega, but it operates in the 30- to 300-kHz low frequency band.

18.2 USING LORAN C FOR AVL

Loran C in land mobile systems has problems ships at sea do not face. Buildings, terrain, and man-made noise cause propagation delays that introduce positional errors. AVL manufacturers have developed techniques that modify Loran C receivers, allowing the host computer to provide adjustments in position based on local propagation conditions. This

Fig. 18-3 Loran C Transmitters

modification, called differential Loran C, improves position accuracy considerably. In general the AVL system in the vehicle consists of a Loran C receiver with its own antenna (approximately 18 in. long), a microprocessor, and a modem that automatically transfers information to the voice channel of the vehicle's ordinary two-way radio. The Loran C information does not interfere with the voice communications. Besides the vehicle's position, the transmitted information includes an identification number and other pertinent data. All is sent to the control point. A computer at the display center, among other things, changes the Loran C coordinate format to a street-oriented format for display purposes, controls a color display, and provides editing for particular fleet functions. Some manufacturers claim that the overall system's accuracy is one block for 95% of the time in 95% of most coverage areas.

The base station automatically polls all the vehicles, which respond by sending position location to the central dispatching computer. Both the polling and the response are in digital format. The computer display shows the position, direction of travel, and indicates if the vehicle has stopped.

18.3 USING A DEAD RECKONING SYSTEM

Dead reckoning (DR) systems are presently being manufactured for AVL. These use odometers to measure distance and a compass to measure direction. The vehicle's starting point must be known. Errors must be continually corrected or they accumulate. Map matching is one correction method. For example, if the compass gives an incorrect heading showing a turn where there is no cross street, the system corrects the error.

Microprocessors are used to control the system. The position and other information are sent in digital form over the regular two-way radio system to a control point where these may be integrated with a CAD system and are also sent to a display computer that sets up the position of the vehicle on a map.

Combining DR with a few radio sign posts or Loran C corrects errors. This enables a calibration of the DR system so errors do not accumulate. DR has achieved an accuracy of 50 ft in tests.

18.4 LOOKING UP AT AVL SATELLITE SYSTEMS

Satellites in orbit are used for circular trilateration or multilateration systems to locate mobile units. There are two basic services: Radio Determination Satellite Service (RDSS) for civilian mobile users and the Global Positioning System (GPS) for the military, which civilians may use to a limited extent. Both systems have limitations for use in high-rise building areas.

RDSS. The FCC assigned special frequency bands to the new RDSS in 1985. The uplink band to the satellite is 6525 MHz with a 16-MHz bandwidth. The downlink from the satellites is 5117 to 5183 MHz with a 64-MHz bandwidth.

GPS. GPS is based on 24 satellites orbiting the earth at 11,000 miles above it. The Department of Defense operates GPS to obtain accurate military navigation. The satellites are not in geo-synchronous orbit. They go around the earth once every 12 hours. The GPS satellites pass over one of the Department of Defense's monitoring stations, which transmits its altitude, position, and speed to the satellites. The satellites transmit a pseudo-random code to a GPS receiver for timing purposes and a data message giving the satellite position.

The system is a three-dimensional one that can be used to locate aircraft, ships and land vehicles. Basically the small GPS receiver measures the distance from it to a satellite by measuring the time it takes for the radio signal to go from the satellite to the GPS receiver. Since distance = time x 186,000 miles/second, the distance from the satellite to the receiver can be computed readily.

If the distance is 20,000 miles from satellite A, then the GPS receiver is located on a sphere that has a radius of 20,000 miles and a center at satellite A. Another satellite B whose distance is measured at 15,000 miles gives a sphere with a radius of 15,000 miles centered on satellite B. The two spheres intersect, forming a circle C. If the GPS receiver is at sea level on the surface of the earth, then there is a third sphere centered at the earth's center and it has a radius equal to the Earth's radius at that point. The third sphere intersects the circle C at two points. One of the two points is ridiculous—for example, it shows an impossibly high velocity—and can be rejected. A tiny computer in the GPS receiver rejects the superfluous solution.

Theoretically three spheres (two satellites for sea GPS receiver locations) can give the vehicle location. However, in practice an additional measurement is required because of the time-keeping devices used.

The satellites use atomic clocks (see section 5.4) as references. They are extremely accurate, but they are too heavy and expensive to use in a small GPS receiver. The GPS receiver uses a small, precise, inexpensive electronic clock. However, its accuracy is one in 10^{-9} seconds while the atomic clock has an accuracy of one in 10^{-12} seconds. Ordinarily this would mean a very large error in position location. An additional satellite measurement is required to give an offset time (or error in timing). The GPS receiver computer can find this offset time by setting up four equations with four unknowns. The four equations come from the four sphere measurements. The computer then can solve for the error due to the timing accuracy difference. Three satellites are required for sea vehicles because the fourth sphere is the earth's sphere.

One basic problem remains: how to determine the time when the timing message from the satellite starts. Both the satellites and the receivers generate a complicated set of digital codes consisting of narrow and wide pulses. These narrow and wide pulses' distribution varies in a particular way that makes up that code. Codes of this type are called pseudo-random codes. The pseudo-random sequences repeat every millisecond. One type of pseudo-random code is used for the military and another is used for civilian use. The satellites use atomic clocks to time the generation of their pseudo-random code. The GPS receivers use less accurate electronic clocks to time their pseudo-random code but are compensated as previously described. The pseudo-random code from the satellite and the pseudo-random code from the GPS receiver are the same. Comparing the satellite pseudo-random code to the pseudo-random code generated in the GPS receiver determines the time difference from the satellite to the receiver.

GPS equipment is now available to be used with land mobile digital radio equipment to send vehicle location data to a dispatching center. Sometimes GPS equipment is combined with other systems such as DR and Loran C to compensate for GPS dead spots caused by buildings blocking satellite signal reception.

REFERENCES

Fey, R. L. "Automatic Vehicle Location Technique for Law Enforcement Use," Electromagnetics Division, National Bureau of Standards. Published by U.S. Department of Justice, Law Enforcement Assistance Administration, National Institute of Law Enforcement and Criminal Justice, Washington, DC; September 1974.

Getting, Ivan A. "The Global Positioning System," *IEE Spectrum*. December 1993.

Gibbs, John. "Vehicle Location in Public Safety," *Communications*. August 1986.

Hern, Jeff "GPS a Guide to the Next Utility." Published by Trimble Navigation, 645 North Mary Avenue, P.O. Box 3642, Sunnyvale, CA 94088–3642; 1989.

Janc, Bob. "The Landsmart AVL System," *Telocator*. August 1983.

Provenzano, Jack. "Automatic Vehicle Location Techniques," *APCO Bulletin, Journal of Public Safety Communications*. January 1987.

Rothblatt, Martin A. *Radiodetermination Satellite Services and Standards*. Dedham, MA: Artech House, Inc., 1987.

Staras, Harold, and Stephen N. Honickman. "The Accuracy of Vehicle Location by Trilateration in a Dense Urban Environment," *IEEE Transactions on Vehicular Technology*. February 1972.

19

Using Satellites for Land Mobile Communication Systems

The FCC has established a mobile satellite service (MSS) with specific frequencies to provide mobile communications for domestic land, air, and marine vehicles. Satellite systems are being designed and constructed to provide voice and data communications for U.S. domestic mobile radio systems.

This chapter reviews the history of experimental satellite land mobile communication systems. Then the advantages and disadvantages of satellite mobile communication systems are compared. The present frequency band assignments in the MSS and their letter designations are listed. The elements of a system for land mobile radio communications are examined. A method of using two satellites to supply both communications and AVL is described. The use of satellites in a railroad system is examined. Finally, different types of satellite systems that may be used in the future land mobile telecommunication system are described.

19.1 THE HISTORY OF SATELLITE MOBILE COMMUNICATIONS

The National Aeronautical Space Administration (NASA) launched a series of experimental advanced technology satellites (ATS) that established the basis for the MSS:

ATS-3

The ATS-3 satellite was launched into a geostationary orbit on November 5, 1967. A geostationary orbit is 22,300 miles above the earth and in the equator's plane. At this height the satellite orbits the earth in the same time that the earth makes one rotation. Since the satellite is traveling in the equator's plane, it appears to be stationary over a specific position.

A transponder aboard the satellite relayed signals between a fixed station on earth and a terrestrial mobile unit. The uplink from the earth to the satellite operated at approximately 150 MHz, while the downlink from the satellite to earth operated at approximately 135 MHz. Circular polarization was used for all antennas.

Using ATS-3. In 1970, Alaskan villages used the ATS-3 to communicate with each other. In 1980, the ATS-3 was used to relay emergency traffic out of the Mount Saint Helena volcano eruption disaster area. A Jeep equipped with special communications equipment sent vital information via ATS-3 to a GE earth station in Schenectady, New York. Here the messages were patched into the telephone system. Federal law enforcement officers also used ATS-3 to send and receive typed messages with three base stations in the United States by relay through the satellite. The federal officers used portable radios and a keyboard and display unit powered either from an automobile cigarette lighter socket or a 115-V ac outlet. GE developed this system under contract to NASA.

Transit Communications was another experimenter with ATS-3. In 1984 this company set up a fixed station in Santa Maria, California and a mobile station operating in the continental United States. Experiments were conducted testing ACSB and other technologies. Today ATS-6 has replaced ATS-3.

ATS-6. In 1972 the state of Alaska started to use the ATS-6 satellite for two-way video as well as audio between remote areas. Over the years 250 earth stations were built.

GE also used this satellite for experiments under contract to NASA. Standard 800-MHz two-way radios in five trucks were modified to operate in the ATS-6 satellite's 1600-MHz L band. The trucks communicated to the GE earth station laboratory near Schenectady via the satellite. Then the message went back up to the satellite, where it was relayed to the base station of the trucking company's dispatch headquarters in Staunton, Virginia. Recordings of the satellite communications were made at the GE earth station. Good signals were received without any of the multipath fading (Rayleigh fading) found in ordinary communications.

OmniTracs. This system developed in 1985 by Qualcomm Corp. uses existing commercial satellites to provide low-speed, two-way data and positioning services for trucking companies.

19.2 COMPARING ADVANTAGES AND DISADVANTAGES OF SATELLITE COMMUNICATIONS

Advantages

There are a number of advantages:

1. *Rural area coverage.* Cellular mobile telephone systems are difficult to implement in rural areas because of the economics. Here satellites may be the answer. Also, rural

areas such as Alaska and other large areas are difficult for ordinary public safety communications to cover. The satellite appears to be the solution in this case.

2. *Eliminates multipath fading in most areas.* A moving vehicle encounters multipath fading when receiving a terrestrial base station. In most areas satellite communications would eliminate multipath fading.

3. *Eliminates co-channel interference.* Co-channel interference can be a serious problem in conventional land mobile radio systems. Satellite relay communication does not have this problem.

4. *Combines AVL with satellite communications.* Satellites in the proper position can combine AVL with satellite communications. (This is discussed in Section 19.5.)

Disadvantages

There are a number of disadvantages:

1. *High-rise building areas.* Satellite communications do not work well in high-rise building areas.

2. *Overpass blockage.* Road overpasses block transmission between mobiles and satellite.

3. *Time lag.* There can be approximately a 0.6-second time lag in the signal between mobile and satellite. This may cause some inconvenience in a two-way voice communication.

General comparison. In general satellite communication is most useful in rural or suburban areas, while terrestrial cellular communication is better in urban areas.

19.3 ASSIGNING FREQUENCIES IN MSS

Literature and advertisements often refer to frequency bands in MSS by letter. Table 19-1 shows the common letter frequency bands in the United States.

TABLE 19-1 COMMON MSS FREQUENCY LETTERS

Band	Frequency (MHz)
L	1,000–2,000
S	2,000–4,000
C	4,000–8,000
X	8,000–12,000
K_u	12,000–18,000
K	18,000–27,000
K_q	27,000–40,000

MSS frequencies between mobile units and satellites in the L band.
The uplink frequencies (from mobile unit to satellite) are 1646.5 to 1660.5 MHz. The downlink frequencies (from satellite to mobile unit) are 1545 to 1559 MHz.

MSS frequencies between gateways (earth stations) and satellites in K_u band.
The communications from the gateway to the satellite are from 14 to 14.5 gigahertz (GHz). The communications from the satellite to the gateway are from 11.7 to 12.2 GHz.

Satellite footprints.
Satellite footprints is a circular or shaped beam from space to mobile units. There may be a large number of footprints, with specific frequencies. This can be used in a satellite-based mobile phone system.

19.4 EXAMINING THE ELEMENTS OF A MOBILE COMMUNICATIONS SATELLITE SYSTEM

Figure 19-1 shows a system that Mobile Satellite Corporation originally proposed.[*] Two satellites are used to relay messages between mobile units and two types of gateways. One type of gateway at a telephone company exchange enables the mobile unit to communicate with anyone using the regular telephone system. The other gateway is a control system that enables a dispatcher to communicate with the mobile unit. The network operating center (NOC) monitors and controls the communications network via RF links. The NOC receives various monitoring data and requests for service over dedicated common signaling channels. Using the common signaling channels, the NOC assigns complementary frequencies to mobiles and gateways. The NOC also monitors the network to assure regulation conformance, performs network diagnostic functions, keeps billing records, and provides operator assistance, if necessary. The NOC addresses the mobiles one by one to introduce commands and instructions. The NOC computes mobiles' position locations. A satellite operating center monitors the condition of the satellite on-board equipment plus satellite attitude and orbital position. The center may be co-located with the NOC and share personnel and computers with the NOC.

The satellite system uses narrowband analog modulation single channel per carrier techniques for 5-kHz channel spacing of voice channels. The two satellites with a 14-MHz-wide allocation provide the 5600 duplex voice channels.

Satellites.
Two satellites are deployed in geostationary orbits 22,300 miles above the equator. Each satellite has an aggregate transmitter power of 2000 W with an antenna of 10 ft. Circular polarization is used for transmission and reception. The type of circular polarization is compatible with the Aviation Mobile Satellite Service, which is being incorporated into MSS.

---

[*] Mobile Satellite Corporation is now part of a consortium, American Mobile Satellite Corp. (AMS). AMS is scheduled to provide service in 1994.

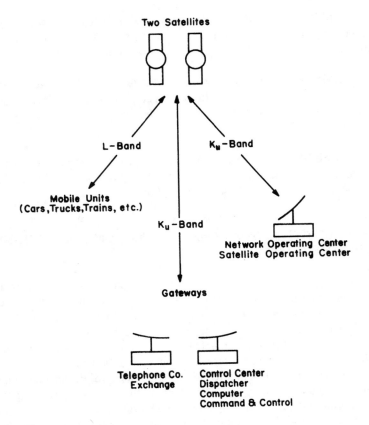

Fig. 19-1 Mobilsat Communications System

Circular Polarization. Circular polarization may be either right-handed or left-handed. In a right-handed circularly polarized wave the electric field rotates in a clockwise direction as seen from the transmitter. In a left-handed circularly polarized wave the electric field rotates in a counterclockwise direction looking from the transmitter. In both cases the direction of rotation is reversed from the receiver's point of view.

Gateway terminals. Two basic types of gateways are used. One, located at or near telephone exchanges, is for radio telephone services. This enables a mobile unit to communicate via the satellite relay with anyone using the regular telephone system. The other type of gateway is a private service at a convenient location. This gateway includes a dispatcher and a central computer. A gateway consists of a transistor receiver, a transmitter rated in the tens of watts, and a 3-m antenna.

Gateway communication with the satellite is in the K_u band. Transmission is on specific frequencies in the 14,000- to 14,500-MHz frequency range. The receiver frequencies are in the 11,700- to 12,200-MHz band. The network operating center automatically assigns on demand, and frequency-division multiplexing is available. The gateways employ an ortho mode transducer (OMT) to select the desired circular polarization.

There are different modems in gateways that are selected according to the type of service used: voice, data, or interactive data/surveillance (IDS). Each uses a specific modem.

Interactive data/surveillance. IDS is a computer-controlled packet protocol used for transferring packets of information between mobiles and gateways. Tens of thousands of IDS mobiles can share a single voice channel. The packet information may consist of selected switch closure control, routing instructions, general information to mobiles, keyboard information, or surveillance data relating to mobiles' position location. The NOC controls the network routing of packets between the NOC, mobiles, and gateways. One end of the IDS link always terminates at the NOC.

Cellular-Compatible mobile radio. A cellular mobile has two antennas, one for terrestrial conventional radio and the other for communications with the satellite. A cellular-compatible mobile radio eventually will be available but probably will not be used in the satellite system's initial operation.

Mobile for satellite communication only. Mobiles for satellite communication only have one antenna and are used for general radio telephone services through the satellite to a teleco gateway. ACSB equipment is used along with a circuit that locks the frequency synthesizer reference to the satellite common signaling channel carrier frequency.

IDS transceiver. Among other functions the IDS transceiver handles information on position location to the NOC via the satellite. The IDS transceiver typically consists of a fixed dual polarized antenna, a fixed channel transceiver, a MODEM/CODEC,* and a memory register with a capacity around one kilobit. Normally the transceiver has only the receiver and decoder on. When the decoder receives a mobile's unique address, the entire transceiver is activated. The mobile unit then may receive instructions to display or to transmit the stored information in its register. The mobile may answer a surveillance command by retransmitting the incoming burst with the mobile identification. This information is transmitted through both satellites to the NOC, where the mobile position is computed. The IDS mobile unit has an OMT to select the desired type of circular polarization.

Mobile antennas. The simplest mobile antenna is a fixed turnstile antenna. This consists of two crossed dipoles with the two elements phased for circular polarization. The elements are bent or curved to maintain circular polarization over the usable beam width. The dimensions are approximately 3.5 by 3.5 in. for L band.

The steerable Yagi antenna is a higher-performance, more expensive antenna. It is mounted on a motor that rotates the antenna in azimuth to point to the satellite. To obtain automatic pointing the common signaling channel carrier level is sensed. The dimensions of a steerable antenna at L band are approximately 4.5 in. high by 12 in. in diameter.

---

* CODEC is an acronym for "coder-decoder."

19.5 COMBINING COMMUNICATIONS WITH AVL IN A SATELLITE SYSTEM

Two satellites can be used for AVL as well as communications. Timing the transmission between a mobile and a satellite defines a circle with the mobile position on the circumference. Each of the two satellites produces a circle that intersects the other circle at two points. The mobile's position is at one of the two intersections. Locating the two satellites so that one intersection is in the northern hemisphere and the other is in the southern hemisphere may resolve this ambiguity, as figure 19-2 illustrates. Knowing the mobile is in the northern hemisphere enables the system to locate the mobile. Accuracy has been established to be within 0.2 miles (about 1000 ft).

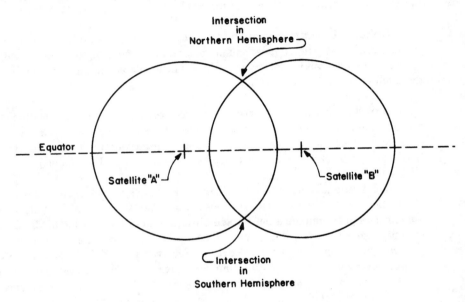

Fig. 19-2 Mobile Position Location Using Two Satellites

19.6 USING SATELLITES FOR RAILROADS

The Santa Fe Railway in the United States now is using satellites to provide centralized traffic control (CTC) service to connect the field control point to a geographically distant controlling center. Trackside pole lines and microwave relays have maintained information from signal computer systems to remote trackside stations in the past. A satellite communication system using very small aperture terminal (VSAT) units at trackside locations is replacing this.

Satellite communications is accomplished between two or more locations via the satellite and a terrestrial HUB station. The HUB is a master station controller, network analyzer, and a matrix switch. It connects the various field stations by supervising the many frequencies and protocols available for traffic and acts as the brains of the system in general.

The CTC computers communicate with the HUB via satellites operating in the K_u band. The HUB communicates with the field stations by satellite on separate frequencies within the same K_u band. The HUB to field operates at 512 KB/s and the field to HUB operates at 128 KB/s. Both use TDMA, allowing a large number of field stations to use the same channel by bursting in a designated time slot, decreasing the cost per site for satellite spectrum charges.

The remote field VSAT. The VSAT consists of two major components: an indoor digital unit and a radio frequency transceiver. A single coaxial cable connects the two components. The transceiver is composed of an offset-fed parabolic reflector antenna and an outdoor unit. The antenna ranges in size from 0.75 to 2.4 meters and has a velox coating to shed water. The outdoor unit is mounted on the antenna feed support and consists of a 1.5- to 3-watt transmitter and a receiver with solid-state low-noise amplifier and block up/down converter. A coaxial cable up to 500 ft connects the outdoor unit to an indoor digital unit that contains six circuit cards. These cards control frequency synthesis, modulation/demodulation program control, and synchronization.

19.7 SATELLITES AND FUTURE PUBLIC LAND MOBILE TELECOMMUNICATION SYSTEMS

Future Public Land Mobile Telecommunication System, or FPLMTS, (the official name designated by the International Telecommunications Union) is intended to become a world-wide personal communications network offering all of a telephone network's services including voice, facsimile, and data. FPLMTS includes cordless telephony, wireless pay phones, private branch exchanges, and rural radio and telephone exchanges among terminals on land, on sea, and in the air. Calls within the mobile system would be routed to and from the existing public switched telephone networks through satellites. FPLMTS will permit connection to the system operating in the mobile-satellite services band. Personal and mobile stations will be able to connect into the system by radio at any time and from any place in the world.

A number of different types of satellite systems can be used in FPLMTS, such as little low-level Earth orbit (LEO), big LEO, and geostationary orbit (GSO).

Little LEO satellites. Little LEO systems provide low-cost, low-data-rate (up to 10 kb/s), two-way digital communications and position-location services to pocket-sized portable and mobile terminals at frequencies below 1 GHz. Such systems would employ two to 24 "lightsats," small inexpensive satellites launched into orbits 480–780 miles above the earth that are capable of serving the world.

Big LEO satellites. A big LEO satellite systems consists of a large number of satellites in low Earth orbits operating above 1 GHz. Big LEO systems operate with traditional voice and a high-speed data service up to a few megabits/second. The big LEO systems would also provide worldwide services in personal communications to handheld terminals like those used in cellular mobile systems.

Motorola Satellite Communications, Inc. has proposed a big LEO system consisting of 77 satellites in low Earth orbit. The company calls its system Iridium[*] after the element whose atom has 77 orbiting electrons.

GSO systems.   Some companies have proposed three satellites in geostationary orbit, each with many spot beams. Hughes Aircraft Co. proposed such a system called Tritium, named after an isotope of hydrogen.

REFERENCES

Anderson, R. E. "Satellites for Public Safety Communications," *APCO Bulletin, Journal of Public Safety Communications.* January 1984.

_____ . "Satellites Meet Rural Needs," *Telephone Engineer and Management.* March 1984.

Bowden, Donald R. and Bowden, Richard P. "Santa Fe Uses Satellite Link for CTC Control," *IEEE Vehicular Technology Society.* Quarterly, February 1993.

Elbert, Bruce, R. *Introduction to Satellite Communication.* Dedham, MA: Artech House, Inc., 1987.

Leung, Victor. "Mobile Radio Group Communications by Satellite," *IEEE Transactions on Vehicular Technology.* Volume 42, No. 2. May 1993.

Pierce, W. "Mobile Satellite Offers Alaska Statewide Coverage," *Mobile Radio Technology.* March 1986.

Peter, Timothy and Root, Don. "Public Safety & Satellite Communications Technology," *APCO Bulletin, Journal of Public Safety Communications.* May 1992.

Pratt, Timothy and Charles W. Bostian. *Satellite Communications.* New York: John Wiley & Sons, Inc., 1986.

Reinhart, Edward E. "Mobile Communications," *IEEE Spectrum, A Special Preview Report.* WARC '92, February 1992.

---

[*] Motorola now plans to use 66 satellites with Iridium service beginning in 1998.

20

Obtaining Station Licenses and Frequency Coordination

This chapter covers station licensing and frequency coordination for private land radio services. License terms including special FCC codes are defined and explained. The required forms, documentation, and how to obtain them are listed.

Frequency coordination is now an official part of the licensing procedure. Each radio service is listed in this chapter, with its frequency coordination agency or agencies. An engineering basis for frequency coordination to avoid co-channel interference is explained. The licensing procedure changes in the proposed FCC Part 88 are discussed.

20.1 DEFINING LICENSING TERMINOLOGY

The FCC uses the following terms and abbreviations for station license applications:

Station Class. Each class of stations has an FCC code for the station license application. Table 20-1 shows the codes and their definitions.

TABLE 20-1 CLASSES OF STATIONS

Station Class	Code	FCC Definitions
Base	FB	A station at a specified site authorized to communicate with mobile stations.
Mobile Relay	FB2	A base station in the mobile service authorized to retransmit automatically on mobile service frequency communications that originate on the mobile station's transmitting frequency.
Community Repeater	FB4	A mobile relay station shared by a group of licensees.

TABLE 20-1 CLASSES OF STATIONS

Station Class	Code	FCC Definitions
Control Station	FX1	An operational fixed station whose transmissions are used to control automatically the emissions or operation of another radio station at a specified location.
Mobile	MO	A station in the mobile service intended to be used while in motion or during halts at unspecified points; includes hand-carried transmitters.
Mobile/Vehicular Repeater	MO3	A mobile station authorized to retransmit automatically on a mobile service frequency communication to or from hand-carried transmitters.
Operational Fixed	FXO	A fixed station, not open to public correspondence, operated by and for the sole use of those agencies operating their own radio communication facilities in the public safety, industrial land transportation, marine, or aviation services.
Fixed Relay	FX2	A station at a specified site used to communicate with a station at another specified site.
Interzone	FXY	A fixed station in the police radio service using radio telegraphy (A-1 emission) for communications with interzone stations in other zones and with stations within the same zone.
Zone	FXZ	A fixed station in the police radio service using radio telegraphy (A-1 emission) for communications with other similar stations in the same zone and with an interzone station in the same zone.
Radiolocation Land	LR	Basic station employing radio determination used for purposes other than radio navigation.
Radiolocation Mobile	MR	Mobile station employing radio determination used for purposes other than radio navigation.

Radio Service Codes. The radio service codes are divided into three parts. Part one covers all frequencies except the 800- and 900-MHz bands. Part two is for frequencies in the 800-MHz band and part three is for the 900-MHz band. Table 20-2 shows the codes for the different radio services.

Output Transmitter Power. For all operations other than single sideband, mean RF output power the transmitter supplies to the antenna feedline is used (in watts). For single-sideband operation, the peak envelope power is used for output power.

Effective Radiated Power (ERP). ERP is the transmitter output power times the net gain of the antenna system. The net gain of the antenna system is the gain of the antenna minus transmission losses, which include losses of transmission line, duplexers, cavity filters, and isolators.

TABLE 20-2 NAMES AND CODES OF THE RADIO SERVICES

Part 1: All Frequencies Except 800- and 900-MHz Bands

Category	Code
Industrial Radio Services	
Business	IB
Forest Products	IF
Motion Picture	IM
Petroleum	IP
Special Industrial	IS
Telephone Maintenance	IT
Power	IW
Manufacturers	IX
Relay Press	IY
Land Transportation Radio Services	
Automobile Emergency	LA
Railroad	LR
Taxicab	LX
Motor Carrier Radio Services	
Interurban Passenger	LI
Interurban Property	LJ
Urban Passenger	LU
Urban Property	LV
Public Safety Radio Services	
Fire	PF
Highway Maintenance	PH
Local Government	PL
Police	PP
Forestry Conservation	PO
Special Emergency	PS
Radiolocation	RS
General Mobile	ZA

Part 2: 806–821/851–866-MHz Bands

Category	Code Conventional	Trunked
Business	GB	YB
Industrial/Land Transportation	GO	YO
Public Safety/Special Emergency	GP	YP
Commercial (SMRS)	GX	YX

Part 3: 929–930-MHz Band

Category	Code
Private Carrier Paging Systems	GS

All other applicants in this frequency band use the code for the radio service in which eligibility is claimed (see Part 1).

Sec. 20.1 Defining Licensing Terminology

Ground Elevation at Antenna Site.   Ground elevation at antenna site is the elevation to the nearest foot above mean sea level of the ground at the antenna location. The elevation can be obtained from a U.S. Geological Survey 7.5-minute topographical map.

Antenna Height to Tip.   Antenna height to tip is the overall height above ground to the nearest foot.

Antenna Height Above Mean Sea Level.   The antenna height above mean sea level is obtained by adding the antenna height above ground to the antenna site's ground elevation.

Antenna Height Above Average Terrain (AAT).   AAT is required for operations in the 470–512-MHz, 806–886-MHz, and 929–930-MHz bands. The elevation of the average terrain above mean sea level is calculated for distances from two to 10 miles from the antenna site. 90.309(a)(4) of the FCC regulations describes the calculations. However, NABER calculates the elevation of the average terrain above mean sea level for a small fee. NABER's address is: National Association for Business and Educational Radio, P.O. Box 19164, Washington, DC 20036. The antenna height above average terrain then is obtained by subtracting the average terrain elevation from the antenna height above mean sea level.

Latitude and Longitude of Antenna Site.   The latitude and longitude are entered in degrees, minutes, and seconds to the nearest second. Accurate latitude and longitude can be obtained from detailed maps in the surveyor's office of the local courthouse, borough hall, or city hall.

Points of Control and Dispatching.   A control point is any place from which a transmitter's functions are controlled. A dispatch point is a place from which radio messages can be originated under the control point's supervision.

Emission Designators.   New emission designators are being used in filling out the FCC Form 574 station license application. The emission designator consists of the necessary bandwidth followed by the emission characteristics. The bandwidth is expressed by three numerals and one letter. The letter occupies the position of the decimal point and represents the unit of bandwidth. For example, a necessary bandwidth of 20 kHz is represented by 20K0.

The emission characteristics consist of a minimum of three symbols. F3E is one commonly used emission characteristic. The first symbol describes the type of modulation; in this case F stands for frequency modulation. The second symbol consists of the nature of signal(s) modulating the main carrier. In this case 3 represents a single channel containing analog information. The third symbol is the type of information to be transmitted. The letter E represents telephony. So 20K0 F3E is the emission designator for a voice-frequency-modulated transmitter with a bandwidth of 20 kHz. Another example used for amplitude compandored single-sideband is 5K00 J3E, where J stands for single-sideband suppressed carrier. Other symbols for emission designators can be found in Sections 2.201 and 2.202 of Part 2 of the FCC regulations.

20.2 OBTAINING THE NECESSARY DOCUMENTATION

There are a number of documents directly or indirectly involved with the licensing process.

Part 2 of the FCC Regulations. Part 2 is included in the Code of Federal Regulations, Telecommunication 47, Parts 0 to 19. Title 47 is revised every October 1 and it can be obtained from the Superintendent of Documents, U.S. Government Printing Office, Washington, D.C. 20402.

Part 90 of the FCC Regulations. Part 90, Private Land Mobile Radio Services, is included in the Code of Federal Regulations, Telecommunication 47, Part 80 to end.

License Application Form 574 and Form 574 instructions. Both the 574 application and instructions can be obtained from a local FCC office. The *FCC Form 574 Instructions* booklet gives information on filling out each item on the license application. The dates on the booklet should cover the date of the license application.

Federal Aviation Administration (FAA) Form 7460–1. FAA Form 7460–1 is a notice of construction or alteration of an antenna tower. It can be obtained from a local FAA office.

Topographical 7.5 minute map. The topographical 7.5 minute map in the area of interest can be obtained from the U.S. Geological Survey, Box 25286, Denver Federal Center, Denver, CO 80225.

Form requesting frequency coordination. The form to request frequency coordination may be obtained from an agency responsible for frequency coordination in your particular radio service. These agencies are listed in section 20.3, "Finding Your Frequency Coordinator."

After the request for frequency coordination is completed, it is sent along with the FCC Form 574 and any other necessary forms to the frequency coordinating agency. When approved these are forwarded to the Federal Communications Commission, Gettysburg, PA 17325 for final approval and license issuance.

20.3 FINDING YOUR FREQUENCY COORDINATOR

Table 20-3 lists the frequency coordinating agency for each radio service. In some cases, such as fire departments, there are two separate frequency coordinators; below 800 MHz there is a separate fire radio service whose frequencies the IMSA coordinates. Above 800 MHz fire radio is incorporated into a public safety/special emergency category, whose frequency coordinating agency is APCO. This same situation is true for forestry conservation and highway maintenance.

TABLE 20-3 FREQUENCY COORDINATING AGENCIES

Radio Service	Frequency Coordinator
Automobile Emergency	American Automobile Association (AAA) National Emergency Road Service 1000 AAA Drive, Mailspace 15 City of Heathrow, FL 407-444-7000 Fax: 407-444-7380
Business	National Association of Business and Educational Radio (NABER) 1501 Duke Street, Suite 200 Alexandria, VA 22314 703-739-0300 Fax: 703-836-1608 NABER does not coordinate frequencies in the 800-MHz trunked frequencies listed in Section 90.362 of the Rules and 800 MHz pool frequencies listed in Sections 90.617(d) and 90.619.
Fire Below 800 MHz	International Municipal Signal Association (IMSA) (jointly with International Association of Fire Chiefs) Alfred J. Mello Frequency Coordination P.O. Box 1513 Providence, RI 02901 401-738-2220 Fax: 315-331-8205
Fire 800-MHz Band	Associated Public Safety Communications Officers, Inc. (APCO) 2040 South Ridgewood Avenue, Suite 202 South Daytona, FL 32119 800-949-2726 Fax: 904-322-2502
Forestry Conservation Below 800 MHz	Forestry Conservation Communications Association (FCCA) 444 North Capitol Street, NW, Suite 540 Washington, DC 20001 202-624-5416 Fax: 202-624-5407
Forestry Conservation 800-MHz Band	APCO
Highway Maintenance Below 800 MHz	American Association of State Highway and Transportation Officials (AASHTO) 444 North Capitol Street, NW Washington, DC 20001 202-624-5800 Fax: 202-624-5469

Radio Service	Frequency Coordinator
Highway Maintenance 800 MHz Band	APCO
Local Government	APCO
Forest Products	Forest Industries Telecommunications P.O. Box 5446 3025 Hillyard Street Eugene, OR 97405 503-485-8441 Fax: 503-485-7556
Motion Picture	Industrial Telecommunications Association 1110 North Glebe Road, Suite 500 Arlington, VA 22201-5720 703-528-5115 Fax: 703-524-1074
Motor Carrier	American Trucking Association, Inc. 2200 Mill Road Alexandria, VA 22314 703-838-1730 Fax: 703-683-1934
Petroleum	Industrial Telecommunications Association
Power	Utilities Telecommunication Council 1620 I Street, NW, Suite 515 Washington, DC 20006 202-872-0030 Fax: 202-872-1331
Railroad	Association of American Railroads 50 F. Street, NW Washington, DC 20001 202-639-2217
Relay Press	Industrial Telecommunications Association
Special Emergency Below 800 MHz 800 MHz Band	Industrial Telecommunications Association APCO
Special Industrial	Industrial Telecommunications Association
Taxicab	International Taxicab Association 3849 Farragut Avenue Kensington, MD 20895 301-946-5700 Fax: 301-946-4641
Telephone Maintenance	Industrial Telecommunications Association

20.4 UNDERSTANDING THE ENGINEERING CONSIDERATIONS FOR FREQUENCY COORDINATION

Frequency coordination is part of the licensing process. It is particularly important in police, fire, and special emergency radio services. If the frequency coordinator does not select frequencies correctly, he or she can cause a great deal of destructive radio interference in emergency communications.

The importance of proper frequency coordination.　Two examples of frequency coordination that wreaked havoc: A public safety communication facility had been operating on certain frequencies for more than 20 years. On one occasion those frequencies were frequency-coordinated for another facility 35 miles away. When the second station went on the air, the result was chaos.

The frequency coordination of a mobile relay on a mountaintop is the second example. The repeater's transmitter was assigned a mobile frequency, which destroyed communications in two places: one 85 miles away and the other 125 miles away.

A solid engineering basis for frequency coordination is necessary to avoid disasters such as these. Such an engineering basis is described in the paper, "Engineering Considerations for Frequency Coordination," by Robert L. Gottschalk, listed in the References. (The following material on the definitions of co-channel interference in frequency coordination is based on Gottschalk's paper, also.)

Defining Co-Channel Interference for Frequency Coordination

When defining co-channel interference for frequency coordination the capture ratio is 6 dB for a base-to-base signal and 12 dB for base-to-mobile or mobile-to-base signals.* This means if a base-to-base signal is 6 dB or more above an interfering signal, there is no destructive interference. A base-to-mobile or a mobile-to-base signal 12 dB or more above an undesired signal also prevents destructive interference. Both the desired and undesired signal levels are evaluated in terms of reliability. Ninety percent is the desired signal level for reliability and 10% is the undesired signal for reliability.

There are two basic classes of co-channel interference. The first is when the applicant's and licensee's base transmitter signals compete at the mobile receivers. The second is when the applicant's base and mobile frequencies are reversed; that is, the applicant's base transmitter is assigned a conventional base frequency. The applicant's base transmitter competes with the licensee's mobile transmitter signal at the licensee's base receiver or voting receiver site.

Base-to-mobile co-channel interference.　Figure 20-1 illustrates the first class of co-channel interference. The 90% reliability is a probability degradation in decibels from the median signal value. The median value can be calculated from one of the computer frequency coordinating programs that are in use. One of these computer programs uses a

---

* From the *FCC Report of the Advisory Committee for the Land Mobile Radio Service.* Vol. 2. 1967.

form of modified Egli equation. The 10% reliability figure is the same absolute value as the 90% reliability figure but opposite in sign. S_1 is calculated using the distance D_1 and S_2 is computed with distance D_2. If S_1 exceeds S_2 by 12 dB, there is no destructive interference from the applicant to the licensee. The same analysis can be performed for interference from the licensee to the applicant.

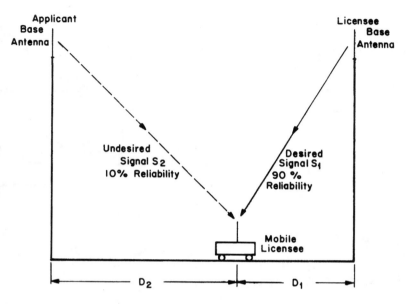

Fig. 20-1 Base-to-Mobile Co-Channel Interference

Co-channel interference, base and mobile frequencies switched. Figure 20-2 illustrates when the applicant files for a license with the base transmitter operating on a regular mobile frequency. The applicant's mobile units transmit on a conventional base frequency. The licensee's mobile transmitter signal S_1 competes with the applicant's base transmitter signal S_2 at the licensee's base receiver. S_1 at a 90% reliability is computed using the licensee's mobile transmitter power. Signal S_2 from the applicant's base transmitter to the licensee's base receiver does not vary with changing location as the mobile does. However, signal S_2 varies somewhat with the seasons, foliage, and so on. The probability factor* for 90% base-to-base line-of-sight communications is 2 dB for 40 MHz, 3 dB for 160 MHz, and 4 dB for 460 MHz. Eight dB is recommended for 90% probability for all bands beyond line-of-sight. These figures can be used for 10% by adding them to the median propagation loss value. If S_1 is 12 dB or more above S_2, the licensee's mobile signal captures the undesired signal from the applicant. If S_2 is 6 dB or more than S_1, the undesired base signal captures the licensee's mobile transmission. This has happened in many situations and is even worse when the licensee has a number of voting receivers.

---

* These probability factors are from Lynch, listed in the References.

Sec. 20.4 Engineering Considerations for Frequency Coordination   **257**

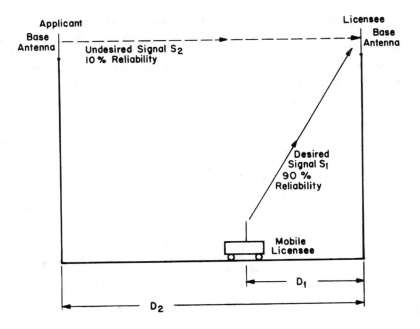

Fig. 20-1 Applicant's Base Transmitter on Mobile Frequency

20.5 REPLACING FCC PART 90 WITH PART 88

The FCC issued a Notice of Proposed Rule Making that was released November 6, 1992 (PR Docket No. 92–235). This concerns replacing Part 90 with Part 88, including modifying licensing procedures in the Private Land Mobile Radio (PLMR) services.

Consolidating radio services. The FCC has proposed two specific alternatives. The first is to consolidate the 10 PLMR services into three broad categories (public safety, non-commercial and specialized mobile radio) plus a general category pool encompassing all three services. The second alternative is to assign the current services to their existing frequency assignments but assign all new frequencies to the proposed new broad categories and the general category pool.

Change in Frequency Coordinators. If the first option is adopted, the FCC proposes that Public Safety Radio Service applicants be permitted to use any of the current public safety frequency coordinators. Non-commercial and general category applicants could use any recognized frequency coordinator. If the second option is adopted, then channels designated for the current 19 radio services would continue to be coordinated only by their current coordinator. Any of the existing coordinators for the Public Safety Radio Services could coordinate channels designated for the Public Safety Radio Service, and any recognized frequency coordinator could coordinate channels designated for the non-commercial radio service and general category pool. APCO, NABER, and the Industrial Tele-communications Association would coordinate the same channels they currently coordinate above 800 MHz. The FCC has stated that in the future frequency coordinators should strive to retain as large a spectrum reserve as possible. For example, frequency recommendations should place systems as geographically close as possible without causing interference.

Reduced limits on effective radio power and AAT. In the 150–174 and 450–470 bands the FCC has proposed reducing the standard limits on effective radio power (ERP) to 300 watts, with lower ERP limits for systems with antenna heights above average terrain greater than 60 meters. This proposal is tied closely to the exclusive use overlay proposal because it would enable the FCC to propose co-channel separations of 50 miles, rather than the 70-mile separation used in the bands above 800 MHz.

Exclusive use overlay (EUO). The FCC proposes permitting exclusive channel assignments in most of the 150–174-MHz, 421–430-MHz, and 450–470-MHz bands. The FCC proposes achieving exclusivity through an exclusive use overlay (EUO) plan. The proposal would permit a temporary licensing freeze on specific channels at specific locations if applicants obtain sufficient concurrence from existing large (as defined by loading criteria) licensees. If concurrence of all large licensees is achieved, the FCC would permanently freeze licensing; that is, no additional use of that channel within 50 miles would be permitted without the EUO licensee's agreement. A special FCC 574X form, *Private Application for Exclusive Use Overlay (EUO) and Temporary Licensing Freeze* will have to be filed to obtain exclusivity.

REFERENCES

Gottschalk, Robert L. "Engineering Considerations for Frequency Coordination." State of Florida, Department of General Services, Division of Communications, Larsen Building, Tallahassee, FL 32301; August 1985.

Lynch, D.J. "Land Mobile Communications Systems Design Guide." State of Florida, Department of General Services, Division of Communications, Larsen Building, Tallahassee, FL 32301; July 1978.

Publications without listed authors

FCC Form 574 Instructions. This can be obtained from the nearest FCC office. Use instructions with dates that include the date of the application.

FCC Report of the Advisory Committee for the Land Mobile Radio Service. Vol. 2. 1967.

Appendix I

The Decibel and Some of Its Disguises

The basic dB term.   The decibel (dB) is defined as 10 log P_1/P_2 where P_1 is the output power and P_2 is the input power. When used in this manner, the answer is positive for gain and negative for loss. If a device's input and output impedances are equal, the decibel can be expressed in terms of output and input voltages:

$$dB = 20 \log V_{out}/V_{in}$$

dBm.   Various references are used to give absolute readings rather than just a ratio. The term dBm uses a milliwatt as a reference:

$$dBm = 10 \log P_1/1 \text{ mW}$$

If P_1 is greater than 1 mW, dBm is a positive value. When P_1 is below 1 mW, the dBm value is negative.

Measuring dBm across an Impedance Different From the Calibrated Meter Impedance.   Many dBm meters are calibrated for measurements across a 600-Ω termination. This meter can be used for any circuit impedance by the following formula:

$$dBm = dBm \text{ indicated} + 10 \log \frac{\text{calibrated impedance}}{\text{circuit impedance}}$$

For example, a meter calibrated for 600 Ω reads 10 dBm across 500 Ω. The corrected reading is

$$10 \text{ dBM} + 10 \log \frac{600}{500} \text{ or } 10.792 \text{ dBm}$$

dBm0.   The 0 refers to 0 TPL, which is the zero transmission-level point. Both the signal and noise levels in a system are referenced to the 0 TPL. For example, a test tone of

1 kHz measured at a –10-dB point in a system has a reading of +25 dBm. Then the dBm0 reading is –10 dB + 25 dB or 15 dBm0. In telephone work the 0 TPL is a reference point on the toll switchboard. In radio work the 0 TPL may be defined as a specific system-level reference point. The 0 ending also is used with other dB readings, such as dBrnc0 and dBa0.

dBrnc.   dBrnc is dB above a reference noise-level using C-message weighting. It is used to show the relative interference effect of two noise power levels: a band of noise frequencies to a reference noise. The reference noise is a 1000-Hz tone at a power level of -90 dBm. The band of noise frequencies used corresponds to a C-message weighting curve. The C-message weighting curve is based on the frequency response of the Western Electric type 500 handset introduced in 1950. A reading of –40 dBm on a meter, with a C-message weighted filter, is 50 dBrnc.

dBa.   dBa is an older unit that measures the noise using Western Electric handset type F1A's frequency response curve. The reference noise is –85 dBm instead of –90 dBm used in dBrnc. A reading of –40 dBm on a meter (FIA weighted) is 45 dBrnc.

Relating dBrnc to dBa.   Some manufacturers use dBrnc and others use dBa, so it is useful to relate one to the other:

$$dBrnc = dBa + 6\ dB \ or\ dBa = dBrnc - 6\ dB$$

Relating Dbrnc to pWp.   The international linear unit psophometric weighting (pWp) is used to measure circuit noise. An approximate conversion to dBrnc is

$$dBrnc = 10\ log\ pWp$$

dBu.   In dBu the dB value is referenced to 1 μV.

dBW.   In dBW the dB value is referenced to 1 W.

Relating dBW to dBm.   0 dBW = 30 dBm.

Relating dBW to Microvolts across 500 Ohms.

$$dBW = 20\ log\ \mu V - 136.99$$

where μ represents microvolts.

Relating dBW to Microvolts Per Meter, Referenced to a Dipole.

$$dBW = 20\ log\left(\frac{\mu V}{m}\Big/F\right) - 105.07$$

where F is the frequency in megahertz and $\frac{\mu V}{m}$ represents microvolts per meter. This is the unit of field strength.

Appendix II

Making Sense of Alphabet Soup (MSOAS!)

AASHTO-American Association of State Highway and Transportation Officials

AAT-Antenna height above average terrain

ACPIR-Adjacent channel protection interference ratio

ACSB-Amplitude-compandored single sideband

ALPC-Adaptive linear predictive coding

ALU-Arithmetic logic unit

AMPS-Advanced mobile phone service

AM-Amplitude modulation

ANI-Automatic number identification

ANSI-American National Standards Institute

ASK-Amplitude shift keying

APCO-Associated Public-Safety Communications Officers, Inc.

ARD-Alarm receipt dispatcher

ARQ-Automatic retry request

ASCII-American Standard Code for Information Interchange

ASPIC-Application specific integrated circuit

AVL-Automatic vehicle location

BARS-Box alarm readout system

BCH-Bose-Chaudhuri-Hocquenghem (error correcting code)

BER-Bit error rate

BICMOS-Bi polar CMOS (integrated circuit combining BIPOLAR and CMOS transistors)

BIT-Binary digit

C4FM-Compatible 4-Level frequency modulation (used in some APCO 25 digital transmitters)

CCIR-Comité Consultatif Internationale Des Radio-Communications (The Consultative International Committee on Radio)

CCITT-Comité Consultatif Internationale Télégraphique et Téléphonique (The Consultative Committee on International Telegraphy and Telephony)

CDMA-Code division multiple access

CELP-Code excited linear predictive (analog-to-digital voice conversion)

CFDD-Compatible frequency discriminator detection (used in APCO 25 digital receivers)

C/N-Carrier-to-noise ratio

CMOS-Complementary metal oxide semiconductor

COMSOC-Communications society

CODEC-COder-DEcoder (used for analog-to-digital conversion)

CP/M-Control program for microprocessors

CPU-Central processing unit

CQPSK-Compatible quadrature phase shift keying (used in some APCO 25 digital transmitters)

CRC-Cyclic redundancy check

CRT-Cathode ray tube

CSMA-Carrier sense multiple access

CTC-Centralized traffic control (trains)

CTCSS-Continuous tone-coded squelch system

CVSD-Continuously variable slope delta modulation

DECT-Digital European cordless telephone system

DD-Decision dispatcher (computer-aided dispatching)

DES-Digital encryption standard

DOD-Department of Defense

DQPSK-Differential quadrature phase shift keying (phase shift reference is position of previous phase shift)

DR-Dead reckoning

DRAM-Dynamic random access memory

DSP-Digital signal processing

DS/SFH-SSMA-Direct sequence/slow frequency hopped spread-spectrum multiple access

DSRS-Digital status reporting system

DS/SSMA-Direct sequence spread-spectrum multiple access

DTMF-Dual-tone multi-frequency

EEPROM-Electrically erasable programmable ROM

EIA-Electronic Industries Association

EIRP-Effective isotropic radiated power

EPROM-Erasable programmable ROM

ERP-Effective radiated power

ERS-Emergency reporting system

ETC-Electronic toll collection

ETI-Emerging technology initiative

ETSI-Institut Européen des Normes de Télécommunications (European Telecommunications Standards Institute)

ETTM-Electronic toll and traffic management

FAMOS-Floating gate avalanche injection metal oxide semiconductor

FCC-Federal Communications Commission

FCCA-Forestry Conservation Communication Association

FDMA-Frequency division multiple access

FDX-Full duplex (data)

FEC-Forward error correction

FET-Field effect transistor

FIFO-First in, first out

FILO-First in, last out

FM-Frequency modulation

FPLMTS-Future public land mobile telecommunication system

FSK-Frequency shift keying

GF-Galois field (used with Reed-Solomon error correction codes)

GIS-Geographical information systems

GMSK-Gaussian minimum shift keying

GPS-Global positioning system

GSC-Golay sequential code

GSM-Group spéciale mobile (European digital cellular system)

GSO-Geostationary orbit

HAAT-Height of antenna above average terrain

HDX-Half duplex (data)

HMOS-High performance metal oxide semiconductor

IDS-Interactive data/surveillance (computer-controlled packet protocol)

IEE-Institution of Electrical Engineers (Great Britain)

IEEE-Institute of Electrical and Electronic Engineers

IF-Intermediate frequency

IM-Intermodulation

IMBE-Improved multiband excitation (vocoder)
IMSA-International Municipal Signal Association
IS-Intermediate standard
ISDN-Integrated services digital network
ISO-International Organization for Standardization
ITU-International Telecommunications Union
IVHS-Intelligent vehicle highway system
LAN-Local area network
LED-Light-emitting diode
LEO-Low-level Earth orbit
LIFO-Last in, first out
LMCC-Land Mobile Communications Council
LORAN-Long-range navigation
mAh-Milliampere ampere hours
MDT-Mobile data terminal
MNOS-Metal nitride oxide semiconductor
MODEM-MOdulator/DEModulator
MOV-Metal oxide varistor
MPP-Multipath propagation
MS-DOS-Microsoft Corporation disk operating system
MSES-Mobile status entry system
MSK-Minimum shift keying
MSS-Mobile satellite service
MTSO-Mobile telecommunications switching office
MUX-Multiplexing
NABER-National Association of Business and Education Radio
NARTE-National Association of Radio and Telecommunication Engineers
NASTD-National Association of State Telecommunications Directors
NBS-National Bureau of Standards
NCS-National Communications Systems
NEC-Nippon Electric Company
NLR-Noise load ratio
NMOS-Negative metal oxide semiconductor
NOC-Network operating center
NPR-Noise Power Ratio
NPSPAC-National Public Safety Planning Advisory Committee
NRZ-Nonreturn to zero
NTIA-National Telecommunications and Information Administration

OMT-Ortho mode transducer (separates polarization in a satellite antenna feed)

OSI-Open systems interconnection (A seven-layer standardized network architecture to enable digital devices to communication with each other)

PAM-Pulse amplitude modulation

PBX-Private branch eXchange

PC-Personal computer

PC-DOS-Personal computer disk operating system

PCM-Pulse-coded modulation

PCN-Personal communications networks

PCS-Personal communication systems

PIA-Peripheral interface adapter

$\pi/4$ DQPSK-Differential quadrature phase shift keying (Phase Shifts are ± 45 degrees and ± 135 degrees)

PIN-Positive-intrinsic-negative (photo diode)

PLMR-Private land mobile radio

PMOS-Positive metal oxide semiconductor

PMR-Private mobile radio

PROM-Programmable read only memory

PSTN-Public switched telephone network

PTT-Push-to-talk

QAM-Quadrature amplitude modulation

QPSK-Quadrature phase shift keying

RACE-European Research Program on third generation cellular system

RAM-Random access memory

RDSS-Radio determination satellite service

RF-Radio frequency

ROM-Read only memory

RS-Reed-Solomon (error correcting code)

SAE-Society of Automotive Engineers

SD-Space diversity

SEP-Status entry panel

SFH-Slow frequency hopping

SINAD-Signal plus noise and distortion to noise plus distortion ratio

SRS-Status reporting system

SMRS-Specialized mobile radio service

SMT-Surface mounting technology

SQL-Structured query language

TAP-Terrain analysis package

TDMA-Time division multiple access
TFM-Tamed frequency modulation
TIA-Telecommunication Industry Association
TNC-Transmitter node controller
TWT-Traveling wave tube
UART-Universal asynchronous receiver/transmitter
UHF-Ultra-high frequency
UMPTS-Universal mobile personal telecommunications services (Europe)
UPS-Uninterrupted power supply
USART-Universal synchronous/asynchronous receiver/transmitter
UTC-Utilities Telecommunication Council
VCO-Voltage controlled oscillator
VHF-Very high frequency
VNIS-Vehicle navigation and information systems
VSAT-Very small aperture terminal (satellite receiver antenna)
VSELP-Vector sum code excited linear predictive (analog to digital voice conversion)
VTS-Vehicular Technology Society
VU-Volume unit
WARC-World administrative radio conference

Index

UHF, 7
UHF-T, 7–8
VHF high-band, 7
VHF low-band, 7
baseband repeaters, 106
base-loaded antenna, 25, 29
base-station antennas, 26–27
    fiberglass radomes, 28
    mounting, 28–29
    wind velocity rating, 30
base-to-mobile co-channel interference, 256–257
batteries, 21–22
bauds, 136
BCH codes, 138
beginner's all-purpose symbolic instruction code (BASIC),
    123
big LEO satellites, 247–248
binary vs. multisymbol systems, 136
bipolar integrated circuits, 127
bits, 136
BNC connectors, 34
Bullington model, 195
bus system, 118–119

C

C language, 123
cable television. See CATV
cables, 32
CADWELD ground connection, 187
carrier sense multiple access (CSMA), 143
carrier squelch, 42–43
carrier-type lines, 94–95
CATV, 112–114
cavity combiners, 57–58
cavity filters, 51–52
cavity-ferrite combiner, 58
CCITTX.25, 145
cellular mobile radio systems, 81–84
    advanced mobile phone service (AMPS) operation,
        82–84
    cell splitting, 82
    co-channel interference, 82
    frequency assignments, 82
Cellular Telecommunications Industry Association, 4
central processing unit (CPU) 117, 119–121
    accumulator, 120
    addresses, 120–121
    ALU, 120
    control and timing logic, 119
    execution, 120–121
    flags, 120
Channel Guard, 45
check sum. See cyclic redundancy check
coaxial collinear antenna, 27
co-channel interference, 208–213
    adding tone-coded squelch, 210–211
    base-to-mobile, 256–257
    direction finder, 212–213

directional antenna, 210, 212
    frequency coordination, 256–258
    New York City Fire Department example, 212–213
    Philadelphia Fire Department example, 208–212
    voting receivers, 208–210
combination duplexer, 55
communication, 197–199
complementary metal-oxide semiconductor (CMOS), 126
computer programs, 202–203
    terrain analysis package (TAP), 203
    U.S. Geological Survey 1:250,000 scale maps, 203
computer to computer via two-way land mobile radio, 143–
    144
computer-aided dispatch (CAD) systems, 13, 130–133
    New York City Fire Department, 131–133
computers, 116–133
    algorithm, 123
    American Standard Code for Information Interchange
        (ASCII), 117
    assembly language, 122
    asynchronous system, 123
    bus system, 118–119
    central processing unit (CPU), 117, 119–121
    complementary metal-oxide semiconductor (CMOS),
        126
    database management systems, 126
    EIA/TIA-232-E standard, 123
    high-level languages, 122
    input/output, 122
    machine language, 122
    memory, 121
    microprocessor word length, 116–117
    microprograms, 123
    mnemonic language, 122
    modems, 122
    n metal-oxide semiconductor (NMOS), 126
    operating systems, 125–126
    p metal-oxide semiconductor (PMOS), 126
    pipeline, 124–125
    random access memory (RAM), 118
    read-only memory (ROM), 117
    synchronous system, 123
    universal asynchronous receiver/transmitter (UART),
        123
    universal synchronous/asynchronous receiver/
        transmitter (USART), 123
conditioned 3002 lines, 95–96
connectors, 34
console control, 13
consoles, 34–38
    digital pulse code, 35
    dual-tone, multifrequency (DTMF), 35
    lights and controls, 36–37
    microphones and headsets, 37–38
    monitoring voting receiver status, 37
    multiple, 38
    mute control, 37
    phone-patch (interconnect), 36
    quick disconnects, 38

Index

frequency shift keying (FSK), 139
frequency spectrum
  amplitude-compandored single sideband, 86
  cellular mobile radio systems, 81–84
  conserving, 78–91
  multiplexing, 89–90
  narrowband digitized voice channels, 88
  other FCC proposals, 91
  proposed time scale, 90–91
  reaching limits of narrowband FM, 87–88
  refarming 90–91
  spread-spectrum techniques, 88–89
  trunking systems, 79–81
frequency synthesizers, 13, 65–67
Fresnel zones, 193
full duplex, 14

G

gain antenna, 25
Global Positioning System (GPS), 237–238
Golay codes, 138
grounding, 187–189
GSO systems, 248
guyed tower, 31

H

half duplex, 14
hamming codes, 138
hand-held portable radios. See portable radios
heterodyne repeaters, 106
high-band VHF (130 to 174 MHz) antennas, 25
high-level languages, 123
high-stability crystal oscillators, 68
hybrid-isolator combiner, 58

I

IEEE Vehicular Technology Society, 6
indirect or subcarrier modulation, 140
industrial radio services, 2
information sources, 6–7
integrated services digital network (ISDN), 135
Intel microprocessors, 129
intermediate frequency (IF), 17
intermodulation, 55–56, 214–218
International Municipal Signal Association (IMSA), 5
international organizations, 7
International Radio Consultative Committee (CCIR), 7
International Taxicab Association, 5
International Telecommunications Union (ITU), 7
International Telegraph and Telephone Consultative
      Committee (CCITT), 7

L

Land Mobile Communications Council (LMCC), 5
land mobile radio systems
  automatic repeat requests (ARQ), 143
  base stations, 141–142
  channels, 143
  combining with computer technology, 116–133
  communications or network processor, 142–143
  contention control, 143
  console control, 13
  digital communications, 13, 140–144
  frequency synthesizers, 13
  half-duplex (HDX) and full-duplex (FDX), 143
  microprocessors, 12, 127–129
  mobile satellite service (MSS), 13
  mobile unit, 141
  polling scheme, 143
  present-day developments, 12–13
  spectrum conservation, 12
  uninterrupted power supply (UPS), 167–168
  vehicles per channel, 143
leaky cable system, 224–227
lightning, 181–191
  damages, 183
  grounding, 184, 187–189
  height of structures and, 183
  land mobile radio systems effects, 182–184
  path of least impedance, 183
  potential on line conductor, 183
  protecting antenna tower, 184–187
  surge protectors, 184, 190–191
line-of-sight microwave transmissions, 99–107
  antennas, 102–103
  combining factors, 100–101
  earth bulge, 99
  equipment-system reliability, 106
  fading, 104–105
  FM modulator and IF stage, 101
  free-space propagation, 104–105
  frequency diversity, 106
  Fresnel phenomenon or diffraction, 100
  hot standby, 106
  limiter and FM discriminator, 103
  local oscillator and down-converter, 103
  microwave equipment, 101–104
  minimum carrier-to-noise ratio (C/N), 104
  multiplexer, 101
  noise, 103–104
  path reliability, 105–106
  preemphasis, 101
  receiver, 103
  refraction of radio waves, 99–100
  repeaters, 106–107
  simplified line path gains and losses, 105
  space diversity, 105–106
  transmitter, 101
  traveling wave tube (TWT), 102
  up-converter, 102

Index

REPEATER AUDIO MONITOR

INCLUDE FACILITY FOR GROUP CALLS — DUMMY SLOTS?

PORTABLE PEP METER